Borbála Fazekas

Computer-assisted enclosures for fourth order elliptic equations

Borbála Fazekas

Computer-assisted enclosures for fourth order elliptic equations

Multiplicity of solutions of nonlinear equations

Südwestdeutscher Verlag für Hochschulschriften

Impressum / Imprint
Bibliografische Information der Deutschen Nationalbibliothek: Die Deutsche Nationalbibliothek verzeichnet diese Publikation in der Deutschen Nationalbibliografie; detaillierte bibliografische Daten sind im Internet über http://dnb.d-nb.de abrufbar.

Bibliographic information published by the Deutsche Nationalbibliothek: The Deutsche Nationalbibliothek lists this publication in the Deutsche Nationalbibliografie; detailed bibliographic data are available in the Internet at http://dnb.d-nb.de.

Coverbild / Cover image: www.ingimage.com

Verlag / Publisher:
Südwestdeutscher Verlag für Hochschulschriften
ist ein Imprint der / is a trademark of
AV Akademikerverlag GmbH & Co. KG
Heinrich-Böcking-Str. 6-8, 66121 Saarbrücken, Deutschland / Germany
Email: info@svh-verlag.de

Herstellung: siehe letzte Seite /
Printed at: see last page
ISBN: 978-3-8381-1834-5

Zugl. / Approved by: Karlsruhe, KIT, Diss., 2012

Contents

Acknowledgements

I wish to express my deep gratitude to my supervisor, Prof. Dr. Michael Plum for the guidance and support during my PhD studies. Without his invaluable help and useful advice this work could not have been completed.

I thank Prof. Dr. Christian Wieners for helping me understand the program M++ and for debugging the errors in the source code.

I express my thanks to Prof. Dr. Wolfgang Reichel for the valuable discussions.

I thank all my former teachers for their lectures of high standard. My special thanks goes to Prof. Dr. Zsolt Páles who deeply influenced me to choose the field of analysis, and encouraged me during my studies.

I would like to thank my parents for my whole education and for their support during my life.

Last but not least I would like to thank my husband for his encouragement and for standing always behind me. I thank my children for their patience with my often not being at home while I was working on my dissertation.

Introduction

Concerning (partial) differential equations, amongst many others two questions are of great importance: existence and uniqueness, or more general multiplicity of solutions. Under certain assumptions we can state existence, or we even have information about the number of the solutions. There are several analytical techniques for proving these properties, such as variational methods or methods based on maximum principles in the second order case. But there are plenty of equations, where analytical methods fail to work. This occurs quite often in the case of higher order equations, since most of the methods working for second order equations can not successfully be applied for the higher order case, and new techniques are available only for a limited number of cases. We summarise the analytical results concerning our problem in Chapter 1.

Even if we can prove existence, we rarely have information about, how the solution looks like. With the aid of numerical methods one can generate numerical solutions. This can help us, to have an idea of the solutions if they exist, but the information, whether a true solution of the equation near to the numerical one exists at all, is usually missing.

The so-called computer-assisted existence proofs can provide all the above information: they prove existence of possibly multiple solutions and they also supply numerical solutions near to the analytical one. The most well-known methods are monotonicity methods, a method developed by Plum (see e.g. [25]), and a method developed by Nakao and his co-authors (see e.g. [17] and [18]). Among the computer-assisted methods the one of Plum fits best our problem, thus we apply this method to obtain existence results. We will discuss these methods in more detail in Chapter 2.

The method of Plum was already successfully applied in many cases, for example for the second order Gelfand-equation (see [26]), or for fourth order travelling waves (in one space dimension, see e.g. [5]) or in many other cases. Our aim is to apply it for two dimensional fourth order equations. As a model problem we chose the fourth order nonlinear biharmonic equation with Dirichlet boundary conditions:

$$\begin{aligned} \Delta^2 u = F(u) \qquad & \text{on } \Omega, \\ u = \frac{\partial u}{\partial \nu} = 0 \qquad & \text{on } \partial\Omega, \end{aligned}$$

where $\Omega \subset \mathbb{R}^2$ is a Lipschitz domain and $u \in H_0^2(\Omega)$.

We will discuss in Chapter 3 how we can apply the method of Plum to the general biharmonic equation.

In the case $F(u) = \lambda \exp(u)$, the equation is called fourth order Gelfand-equation, and in the case $F(u) = \lambda + u^2$, the equation is called fourth order Emden-equation. We will demonstrate in Chapters 4 and 5 respectively with the help of these examples how our method works.

The method of Plum requires a numerical approximation of the solution, in our case in the space $H_0^2(\Omega)$. As we calculate these approximations via finite element method, we would need C^1-finite elements. But compared to C^0-elements C^1-elements are quite difficult to implement, moreover they are numerically expensive. Thus our aim is to use only continuous elements. Therefore we reformulate our problem in a system

$$\begin{aligned} -\Delta u = & \quad v & \text{on } \Omega, \\ -\Delta v = & \quad F(u) & \text{on } \Omega, \\ u = \frac{\partial u}{\partial \nu} = & \quad 0 & \text{on } \partial\Omega. \end{aligned}$$

To obtain numerical approximations to the weak solutions $u \in H_0^1(\Omega)$ and $v \in H^1(\Omega)$ of this system, one only needs C^0-finite elements. We will work numerically with these continuous approximations to u and v. Of course, for the method of Plum we still need a numerical approximation $\omega \in H_0^2(\Omega)$. But we are not going to calculate ω, we will only define it via

$$\Delta^2 \omega = \Delta div\ \sigma,$$

where $\sigma \in (H_0^1(\Omega))^2$ denotes a numerical approximation to ∇u. All the amounts needed for Plum's method will be led back to amounts containing only C^0-approximations. Chapter 3.2 contains these results.

For our computations we will need auxiliary tools such as the so-called div-rot constant, imbedding constants and bounds for given eigenvalues. Furthermore we also make use of some numerical tools, such as interval arithmetics, validated quadrature formulas and homotopy methods (to get enclosures for certain eigenvalues). Chapter 6 contains these results.

1 Analytical results for the biharmonic equation

It is widely known that among partial differential equations second order equations play the most important role in the field of physics. But besides this, fourth order equations also describe basic processes. One can mention from the theory of elasticity the oscillation of an elastic rod, or analogously the transverse vibration of a thin plate. The corresponding partial differential equation is

$$\Delta^2 u = -\frac{1}{c^2}\frac{\partial^2 u}{\partial t^2}$$

in one and two space dimensions, respectively. Here c denotes the velocity of sound in elastic material. Further frequently studied problems of fourth order elliptic equations are for example models for stationary surface diffusion flow, the Paneitz-Branson equation or the stream function formulation of Navier-Stokes problems.

The operator Δ^2 is a prototype of an elliptic operator of order 4, thus the biharmonic equation

$$\Delta^2 u = F(u)$$

is a prototype of a nonlinear fourth order elliptic equation.

The biharmonic equation in two space dimensions arises in continuum mechanics by plates. If force F acts on a plate, then the solution $u(x, y)$ of the biharmonic equation describes the deformation of the plate under the effect of F.

For our model we still need to choose boundary conditions for the equation. For higher order problems a wide class of boundary conditions is available. For the biharmonic equation the most relevant boundary conditions for the field of physics are Dirichlet boundary conditions

$$u = h_1, \qquad \frac{\partial u}{\partial \nu} = h_2,$$

Navier boundary conditions

$$u = h_1, \qquad \Delta u = h_2,$$

and Steklov boundary conditions

$$u = h_1, \qquad \Delta u - a\frac{\partial u}{\partial \nu} = h_2.$$

If we pose Dirichlet boundary conditions, we have the so-called clamped plate problem, with Navier and Steklov boundary conditions we have the hinged plate problem with neglecting or considering the contribution of the curvature of the boundary, respectively. The choice of the boundary condition influences considerably the behaviour of the solution, e.g. the positivity preserving property.

As a model problem we choose the biharmonic equation with homogeneous Dirichlet boundary conditions on a bounded Lipschitz domain.

In the linear case, i.e. if F does not depend on u the Lax-Milgram Lemma ensures the existence and uniqueness of the solution in $H_0^2(\Omega)$. The general theory of boundary value problems for linear elliptic operators of order $2m$ was developed by S. Agmon, A. Douglis and L. Nirenberg, see [1], [2], [3] and [7]. The nonlinear case is much more complex. The recently published book of F. Gazzola, H.-Ch. Grunau and G. Sweers gives a nice detailed summary of present results and of new techniques concerning the general polyharmonic operator, i.e.,

$$(-\Delta)^{2m} + A,$$

where A contains all the lower order derivatives. The authors mostly focus on regularity and positivity preserving properties amongst other ones, and less on the existence and multiplicity of the solutions.

In a recent paper of W. Reichel and T. Weth (see [27]) one finds existence and multiplicity results for the strong solutions of the polyharmonic equation on C^4-domains with certain restrictions on the nonlinearity such as growth-conditions. These results can be applied to the Emden-equation, but they fail for the fast increasing exponential nonlinearity of the Gelfand-equation. Also in case of the Emden-equation existence and multiplicity results are obtained, by the methods of [27] and [4], only for small values of λ, see Subsection 1.1.

Also for small values of λ we could prove existence of two weak solutions of the Emden-equation on Lipschitz domains with standard variational methods, see Subsection 1.1.

Ch. Wieners proved in [30] with numerical methods the existence of one solution for larger values of lambda, more precisely for values of lambda near to the turning point. The proof is based upon the method of Plum, which we will describe in Section 2.2.1.

The author does not have knowledge about other results of existence and multiplicity of the solutions of the Gelfand-equation and of the Emden-equation with λ not near to 0. The aim of the present work is to show existence and multiplicity of weak solutions of the Emden-, and the Gelfand-equations on at most C^1-domains, and if possible for larger values of λ.

1.1 Existence of multiple solutions of the Emden-equation by variational methods

Consider the parameter dependent nonlinear equation on a bounded C^4-domain $\Omega \subset \mathbb{R}^2$

$$Lu = u^2 + \lambda \quad \text{on } \Omega, \tag{1}$$

$$u = \frac{\partial u}{\partial \nu} = 0 \quad \text{on } \partial\Omega,$$

or more general

$$\begin{aligned} Lu &= f(x,u) \quad \text{on } \Omega, \\ u = \frac{\partial u}{\partial \nu} &= 0 \quad \text{on } \partial\Omega, \end{aligned} \tag{2}$$

where $L = \Delta^2$ and $f\colon \Omega \times \mathbb{R} \to \mathbb{R}$.

We are aiming at proving existence of multiple solutions of (1) with the help of variational methods. Let us summarise first a part of the results concerning (2) that are known in the literature.

Paper [27] deals with strong solutions of (2). Let us introduce the following assumptions on f and on L as in the paper:

(H1) $f\colon \Omega\times\mathbb{R} \to \mathbb{R}$ is uniformly continuous in bounded subsets of $\Omega\times\mathbb{R}$ and there exist $q > 1$ and two continuous functions $k, h\colon \overline{\Omega} \to (0,\infty)$ such that

$$\lim_{s\to\infty} \frac{f(x,s)}{s^q} = h(x) \qquad \lim_{s\to-\infty} \frac{f(x,s)}{|s|^q} = k(x)$$

uniformly with respect to $x \in \overline{\Omega}$.

(H2) $f(x,s) = o(s)$ as $s \to 0$ uniformly in $x \in \Omega$.

(H3) For some $p \geq 1$, $v = 0$ is the unique solution of $Lv = 0$ in $H^{4,p}(\Omega) \cap H_0^{2,p}(\Omega)$.

Problem (1) fulfils (H1) with $f(x,u) = u^2 + \lambda$ with $h = k = 1$ and $q = 2$ for all $\lambda \in \mathbb{R}$, (H2) for $\lambda = 0$, and L fulfils (H3). Thus as special cases of Theorems 1, 2 and 3 in [27] we can state the following lemmata:

1.1 Lemma *Problem (1) with $\lambda = 0$ has a nontrivial solution in $H^{4,p}(\overline{\Omega}) \cap H_0^{2,p}(\Omega)$ for all $1 \leq p < \infty$.*

1.2 Lemma *There exists a value $\Lambda \in \mathbb{R}$ such that for $\lambda \geq \Lambda$ problem (1) has no solution in $H^{4,p}(\Omega) \cap H_0^{2,p}(\Omega)$ for all $1 \leq p < \infty$.*

1.3 Lemma *For every compact interval* $[-\Lambda_0, \Lambda]$ *there exists a value* $C > 0$ *such that for all solutions* $u \in H^{4,p}(\Omega) \cap H_0^{2,p}(\Omega)$ *of problem (1) with* $\lambda \in [-\Lambda_0, \Lambda]$

$$\|u\|_\infty \leq C$$

holds for all $1 \leq p < \infty$.

Proof: The constant C in Theorem 1 in [27] depends only on Ω and on q, h, k of (H1), and not directly on the function f itself. As the quantities q, h, k of (H1) do not depend on λ, therefore C does not depend on λ either. Thus the assertion holds. □

If we consider pairs (λ, u_λ), where u_λ is a solution to (1), then we can state more. Beginning with the trivial pair $(0, 0)$ we can continue this pair to a global solution continuum (λ, u_λ) in λ. More precisely under certain assumptions we can prove the existence of two connected, closed sets $C^+ \subset [0, \infty) \times C(\overline{\Omega})$ and $C^- \subset (-\infty, 0] \times C(\overline{\Omega})$ of pairs (λ, u_λ). This statement is an application of Theorem 3.3 in [4] cited next.

1.4 Theorem *Let* X *be a Banach space,* $F\colon \mathbb{R} \times X \to X$ *such that for all* $\lambda \in \mathbb{R}$ *the mapping* $F(\lambda, \cdot)\colon X \to X$ *is compact and* $F(\lambda, x)$ *is continuous in* λ *uniformly with respect to* x *in balls in* X. *Let* (λ_0, x_0) *be a solution of*

$$x = F(\lambda, x). \tag{3}$$

Suppose $U \subset X$ *is an open, bounded set such that* $x_0 \in U$ *and*

(i) *there is no other solution of (3) in* $\overline{U}$ *for* $\lambda = \lambda_0$,

(ii) $\deg(Id - F(\lambda_0, \cdot), U, 0) \neq 0$.

Then there exist two connected and closed sets $C^+ \subset [\lambda_0, \infty) \times X$ *and* $C^- \subset (-\infty, \lambda_0] \times X$ *of solutions of (3) with* $(\lambda_0, x_0) \in C^+ \cap C^-$. *For* C^+ *at least one of the following two statements holds:*

(a) C^+ is unbounded

or

(b) $C^+ \cap (\{\lambda_0\} \times (X \setminus \overline{U})) \neq \emptyset$.

The same holds for C^-.

A summary of the Leray-Schauder degree appearing in (ii) can be found in [4], in [6] or in [19].

To be able to apply Theorem 1.4 we rewrite our problem to a fixed point problem as

$$u = L^{-1}(\lambda + u^2), \tag{4}$$

where $L^{-1}\colon C(\overline{\Omega}) \subset L^p(\Omega) \to H^{4,p}(\Omega) \cap H_0^{2,p}(\Omega)$ denotes the bounded inverse of L, with $p > 1$.

1.5 Corollary *There exist two connected and closed sets $C^+ \subset [0,\infty) \times C(\overline{\Omega})$ and $C^- \subset (-\infty, 0] \times C(\overline{\Omega})$ of solutions of (4) with $(0,0) \in C^+ \cap C^-$.*

Moreover, C^+ is bounded and there exists an $\varepsilon > 0$ such that $C^+ \cap (\{0\} \times (C(\overline{\Omega}) \setminus \overline{B_\varepsilon(0)})) \neq \emptyset$, and furthermore for C^- at least one of the following two statements holds:

(a) C^- is unbounded

or

(b) $C^- \cap (\{0\} \times (C(\overline{\Omega}) \setminus \overline{B_\varepsilon(0)})) \neq \emptyset$.

Proof: We have to show that our problem fulfils the assumptions of Theorem 1.4. Let $X = C(\overline{\Omega})$. Since the imbedding $H^{4,p}(\Omega) \cap H_0^{2,p}(\Omega) \hookrightarrow C(\overline{\Omega})$ is compact, the operator $L^{-1}\colon C(\overline{\Omega}) \to C(\overline{\Omega})$ is compact. The pair $(\lambda_0, u_0) = (0,0)$ is clearly a solution of (4). As in the proof of Theorem 2 in [27] we show that an open ball $B_\varepsilon(0) \subset C(\overline{\Omega})$ exists, such that (i) and (ii) are fulfilled. Due to the compactness of L^{-1} and to (H2), it holds that $\|L^{-1}(u^2)\|_\infty = o(\|u\|_\infty)$

as $\|u\|_\infty \to 0$. Therefore there exists a $0 < \delta < 2$ such that

$$\frac{\|L^{-1}(u^2)\|_\infty}{\|u\|_\infty} < 1/2 \tag{5}$$

holds for $\|u\|_\infty \leq \delta$, $u \neq 0$. Thus $\|tL^{-1}(u^2)\|_\infty < \frac{\delta}{2}$ for $\|u\|_\infty = \delta$ and $t \in [0,1]$. Therefore the operator $\mathrm{Id} - tL^{-1} \circ \omega \colon C(\overline{\Omega}) \to C(\overline{\Omega})$ with $\omega(u) = u^2$ does not attain the value 0 on the boundary of $B_\delta(0)$. Since $\deg(\mathrm{Id} - 0 \cdot L^{-1} \circ \omega, B_\delta(0), 0) = 1$, by the homotopy invariance of the degree we have $\deg(\mathrm{Id} - 1 \cdot L^{-1} \circ \omega, B_\delta(0), 0) = 1 \neq 0$.

From (5) it follows that the only solution of $u = L^{-1}(u^2)$ in $B_\delta(0)$ is $u = 0$. Thus the assumptions of Theorem 1.4 are fulfilled, which ensures the existence of the sets C^+ and C^- and the corresponding statements.

From Lemma 1.2 it follows that the set C^+ is bounded in the λ-direction and from Lemma 1.3 it follows, that it is bounded in the u-direction. Thus for C^+ only case (b) can hold. □

From Corollary 1.5 it follows, that on bounded C^4-domains and for small positive values of λ there exist at least two strong solutions of (1). If we weaken the assumptions on Ω, we can still prove with variational methods for small values of λ the existence of two solutions of (1), now in the space $H_0^2(\Omega)$ equipped with the norm $\|u\|_{H_0^2} = \|\Delta u\|_{L_2}$. For the proof we will use the following version of the Mountain Pass Theorem:

1.6 Theorem (Mountain Pass Theorem) *Let H be a Banach space, $J \in C^1(H, \mathbb{R})$ be a functional and let $u_0 \in H$. If*
(i) there exist $\rho > 0$ and $\alpha \in \mathbb{R}$, such that $J[u] \geq \alpha$ for $\|u\| = \rho$, $\|u_0\| < \rho$, and $\alpha > J[u_0]$
(ii) there exists $v \in H$, such that $\|v\| > \rho$ and $J[v] < J[u_0]$,
(iii) J fulfils the Palais-Smale condition, i.e., every sequence $(u_n)_{n \in \mathbb{N}} \subset H$

such that $(J[u_n])_{n\in\mathbb{N}}$ is bounded and $J'[u_n] \to 0$ in H, has a convergent subsequence in H,
then

$$c = \inf_{g\in\Gamma} \max_{t\in[0,1]} J[g(t)], \quad \textit{where} \quad \Gamma = \{g \in C([0,1], H) \colon g(0) = u_0, g(1) = v\}$$

is a critical value of J.

For the proof we refer to [29].

Our theorem about the multiple solutions of (1) in $H_0^2(\Omega)$ on bounded Lipschitz domains is the following.

1.7 Theorem *Let $\Omega \subset \mathbb{R}^2$ be a bounded Lipschitz domain. Then there exists $\lambda_0 > 0$ such that problem (1) has at least two solutions in $H_0^2(\Omega)$ for $0 \leq \lambda < \lambda_0$.*

Proof: Let us define the functional

$$J[u] = \int_\Omega \frac{(\Delta u)^2}{2} - (\lambda u + \frac{1}{3}u^3)\, dx$$

for $u \in H_0^2(\Omega)$. We are looking for $u \in H_0^2(\Omega)$ such that $J'[u] = 0$. Then u is a weak solution of (1).

First we show, that there exist $\rho > 0$ and $\alpha > 0$ such that $J[u] \geq \alpha$ for all $u \in \partial B_\rho(0)$. For this purpose we estimate $J[u]$ from below as follows:

$$J[u] \geq \int_\Omega \frac{(\Delta u)^2}{2} - \lambda|u| - \frac{1}{3}|u|^3\, dx \geq$$

$$\geq \frac{1}{2}||\Delta u||_{L_2}^2 - \lambda C_{H_0^2 \hookrightarrow L_1} ||\Delta u||_{L_2} - \frac{1}{3} C_{H_0^2 \hookrightarrow L_3}^3 ||\Delta u||_{L_2}^3,$$

where $C_{A\hookrightarrow B}$ denotes the imbedding constant from the space A to the space B. Substituting $||\Delta u||_{L_2} = t$, $a = \frac{1}{3} C_{H_0^2 \hookrightarrow L_3}^3$, $b = C_{H_0^2 \hookrightarrow L_1}$ we have to find $t_0 > 0$

such that $g(t_0) > 0$, where

$$g(t) = \frac{1}{2}t^2 - \lambda C_{H_0^2 \hookrightarrow L_1} t - \frac{1}{3} C^3_{H_0^2 \hookrightarrow L_3} t^3 = t(-at^2 + \frac{1}{2}t - b\lambda).$$

A simple analysis of g shows that in case $\lambda < \frac{1}{16ab}$ the function g has three roots, namely $t_1 = 0$, $0 < t_2 < \frac{1}{4a}$ and $\frac{1}{4a} < t_3$, and g is positive on (t_2, t_3). Thus we can choose for ρ any value in (t_2, t_3), and for α the corresponding function value $g(\rho)$.

The above calculations also show, that J is bounded from below on $B_\rho(0) \subset H_0^2(\Omega)$, thus there exists a minimising sequence of J on $B_\rho(0)$. Let us choose now a minimising sequence $(u_n)_{n\in\mathbb{N}} \subset B_\rho(0)$, such that $J'[u_n] \to 0$, $(n \to \infty)$ also holds. As J is lower semicontinuous, bounded from below on $B_\rho(0)$ and $J[0] = 0 < \alpha \leq J[v]$ for $v \in \partial B_\rho(0)$, the variational principle of Ekeland ensures the existence of such a sequence. Since $(u_n)_{n\in\mathbb{N}}$ is bounded, there exists a weakly convergent subsequence of $(u_n)_{n\in\mathbb{N}}$. W.l.o.g. let $(u_n)_{n\in\mathbb{N}}$ be this subsequence. Let us denote its weak limit by u_0. Then for all $\varphi \in H_0^2(\Omega)$

$$J'[u_n](\varphi) = \int_\Omega \Delta u_n \Delta\varphi - \lambda\varphi - u_n^2\varphi \, dx \to \int_\Omega \Delta u_0 \Delta\varphi - \lambda\varphi - u_0^2\varphi \, dx, \quad (6)$$

since

$$\int_\Omega \Delta u_n \Delta\varphi \to \int_\Omega \Delta u_0 \Delta\varphi$$

due to the weak convergence of $(u_n)_{n\in\mathbb{N}}$ and

$$\left| \int_\Omega (u_n^2 - u_0^2)\varphi \, dx \right| \leq \|u_n + u_0\|_{L_4} \|u_n - u_0\|_{L_4} \|\varphi\|_{L_2}$$

$$\leq 2C_{H_0^2 \hookrightarrow L_4} \rho \|\varphi\|_{L_2} \|u_n - u_0\|_{L_4} \to 0,$$

since $u_n \to u_0$ in L_4 because of the compact imbedding $H_0^2(\Omega) \hookrightarrow L_4(\Omega)$. From $J'[u_n] \to 0$ and (6) it follows that u_0 is a solution of (1) in $H_0^2(\Omega)$.

We can prove the existence of a second solution with the help of the Mountain

Pass Theorem. We verify the assumptions of the lemma.

1. The functional J is clearly continuously differentiable.

2. Let u_0 be the first solution from the above results. Then due to the construction of u_0 and to the lower semicontinuity of J it holds that $J[u_0] = \inf_{u\in B_\rho(0)} J[u] \leq J[0] = 0$. Thus the above ρ and α fulfils assumption (i) of the Mountain Pass Theorem.

3. J fulfils the Palais-Smale compactness condition. To see it let $(u_n)_{n\in\mathbb{N}} \subset H_0^2(\Omega)$ such that $(J[u_n])_{n\in\mathbb{N}}$ is bounded and $J'[u_n] \to 0$ in $H_0^2(\Omega)$. Then there exists a constant $C > 0$ such that

$$C > \left| J[u_n] - \frac{1}{3} J'[u_n](u_n) \right|$$

$$= \left| \int_\Omega \frac{1}{2}(\Delta u_n)^2 - \lambda u_n - \frac{1}{3} u_n^3 \, dx - \int_\Omega \frac{1}{3}(\Delta u_n)^2 - \frac{\lambda}{3} u_n - \frac{1}{3} u_n^3 \, dx \right|$$

$$= \left| \frac{1}{6} \int_\Omega (\Delta u_n)^2 - 4\lambda u_n \, dx \right| \geq \frac{1}{6} \left(\|u_n\|_{H_0^2}^2 - 4\lambda \|u_n\|_{L_1} \right)$$

$$\geq \frac{1}{6} \left(\|u_n\|_{H_0^2}^2 - 4\lambda C_{H_0^2 \hookrightarrow L_2} \|u_n\|_{H_0^2} \right).$$

This means, that $(u_n)_{n\in\mathbb{N}}$ is bounded in $H_0^2(\Omega)$.

Therefore $(u_n)_{n\in\mathbb{N}}$ has a weakly convergent subsequence in $H_0^2(\Omega)$. Let us denote this subsequence also by $(u_n)_{n\in\mathbb{N}}$ and the weak limit by u. Due to the compact imbedding from $H_0^2(\Omega)$ to $L_1(\Omega)$ and $L_3(\Omega)$, the sequence $(u_n)_{n\in\mathbb{N}}$ converges in norm in $L_1(\Omega)$ and in $L_3(\Omega)$

Due to the weak convergence of $(u_n)_{n\in\mathbb{N}}$ it holds that

$$0 \leq \liminf_{n\to\infty} \int_\Omega (\Delta u - \Delta u_n)^2 \, dx = \liminf_{n\to\infty} \int_\Omega (\Delta u)^2 - 2\Delta u \Delta u_n + (\Delta u_n)^2 \, dx$$

$$= \liminf_{n\to\infty} \int_\Omega (\Delta u_n)^2 - (\Delta u)^2 \, dx.$$

Together with the convergence of $(u_n)_{n\in\mathbb{N}}$ in $L_1(\Omega)$ and in $L_3(\Omega)$ this results in

$$\int_\Omega (\Delta u)^2 - \lambda u - u^3\, dx \leq \liminf_{n\to\infty} \int_\Omega (\Delta u_n)^2 - \lambda u_n - u_n^3\, dx = J'[u_n](u_n). \quad (7)$$

Moreover,

$$J'[u_n](u_n) \to 0,$$

since $J'[u_n] \to 0$ and $(u_n)_{n\in\mathbb{N}}$ is bounded. Thus the liminf in (7) exists as a limes and it is equal to 0. Again since $J'[u_n] \to 0$ it holds for all $\varphi \in H_0^2(\Omega)$ that

$$J'[u_n](\varphi) \to 0.$$

On the other hand as in the first part of the proof we can show that

$$J'[u_n](\varphi) \to J'[u](\varphi),$$

therefore we obtain with $\varphi = u$ that

$$0 = J'[u](u) = \int_\Omega (\Delta u)^2 - \lambda u - u^3\, dx.$$

Thus (7) holds with equality, i.e.,

$$\int_\Omega (\Delta u)^2 - \lambda u - u^3\, dx = \lim_{n\to\infty} \int_\Omega (\Delta u_n)^2 - \lambda u_n - u_n^3\, dx.$$

Together with the convergence of $(u_n)_{n\in\mathbb{N}}$ in $L_1(\Omega)$ and in $L_3(\Omega)$, this results that

$$\|u_n\|_{H_0^2} \to \|u\|_{H_0^2}.$$

From the weak convergence of $(u_n)_{n\in\mathbb{N}}$ follows then $u_n \to u$ in $H_0^2(\Omega)$. This means, that J fulfils the Palais-Smale condition.

4. For finding an appropriate v for assumption (ii) let $v_t = t \cdot w$ for $w \in H$

with $\int_\Omega w^3\,dx > 0$. Then

$$J[v_t] = \int_\Omega \frac{t^2}{2}(\Delta w)^2 - t\lambda w - \frac{t^3}{3}w^3\,dx \to -\infty, \quad \text{as} \quad t \to \infty,$$

therefore for sufficiently large t inequality $J[v_t] < J[u_0]$ holds.

Thus all the assumptions of the Mountain Pass Theorem is fulfilled. This means, that

$$c = \inf_{g\in\Gamma}\max_{t\in[0,1]} J[g(t)], \quad \text{where } \Gamma = \left\{g \in C([0,1], H_0^2(\Omega)) \colon g(0) = u_0, g(1) = v\right\}$$

is a critical value of J, i.e., there exists $u_1 \in H_0^2(\Omega)$ such that $J'[u_1] = 0$ and $J[u_1] = c$. Therefore u_1 is a solution of (1), which is clearly different from u_0, since $J[u_0] \leq 0 < \alpha \leq c = J[u_1]$. □

2 Computer-assisted enclosure methods

First of all let us make clear what we mean by computer-assisted proof and enclosure method occurring in the title.

2.1 Computer-assisted proofs

A computer-assisted proof is a mathematical proof that has been at least partially generated by computers. The role of the computer can vary from a very small contribution (e.g. calculating the inverse of a matrix via Maple or Matlab) to the entire proof (automated theorem proving). In case of automated theorem provers the computer works in a given logic with axioms and formulas, and in the favourable case it decides whether the given formula is valid or not. But we do not go that way, since there is no chance to obtain fully automatised proof for nonlinear partial differential equations. We "only" use the computer for getting numerical approximations to some functions, for calculating given integrals (validated), for solving huge eigenvalue problems. If all the numerical approximations are good enough in some sense, then Theorem 2.2 ensures the existence of a true solution of the given problem near to the approximate solution.

It is a controversial question in mathematics whether these proofs are reliable or not. The computer-program as well as the hardware can contain several bugs. A way out can be, if the program is run on several machines. We used 6 different computers to calculate our results. Furthermore, there are also methods to decide about the correctness of certain programs. For a more detailed discussion of reliability we refer to e.g. [11].

2.2 Enclosure methods

Let X be a function space and $u \in X$ be a given function. Enclosure means generally a subset $U \subset X$ which fulfils $u \in U$. One is of course interested in a "small" subset U which then gives us information about the function u itself. In practice the enclosing subset U is e.g. a ball with center u and a given (small) radius, or in case of an ordered space X an interval $\{v \in X : u_1 \leq v \leq u_2\}$ with a small diameter $\|u_1 - u_2\|_X$.

The most well-known enclosure methods are based upon monotonicity principles. Maybe the simplest example in the classical theory of partial differential equations is the monotonicity principle for the equation

$$-\Delta u = F(x, u)$$

with Dirichlet boundary condition. The theorem follows from the maximum principle for subharmonic functions and states that if the right-hand side F is monotone in some sense, then one can construct "closely neighbouring" upper and lower solutions u_1 and u_2, and thus obtain an enclosure $u_1(x) \leq$

$u(x) \leq u_2(x)$ for some solution u.

There are generalisations of this method for more general partial differential equations, relying on the monotonicity of an operator and on a fixed-point argument, see e.g [23] for a summary and references therein.

There is also another enclosure method developed by M.T. Nakao and his coauthors, see e.g. [17] and [18].

The monotonicity methods have the defect that they work only for a limited class of equations. A newer method developed by M. Plum can be applied for a much more general class of problems. It is an existence and enclosure theorem based upon the Schauder fixed point theorem. We will use this method to get our results for fourth order problems. We discuss this method in detail in the next section.

2.2.1 Existence and enclosure method of Plum

In this section we describe the existence and enclosure method of Plum on a general level. In Section 3 we show the application of the general theorem to fourth order problems.

Let the following notations and assumptions hold throughout this chapter. Let X, Y, Z denote three Banach spaces such that $X \subset Y$ and that the imbedding $E_X^Y : X \to Y$ is compact. Let $\mathcal{F} : Y \to Z$ denote a Fréchet differentiable operator. Moreover, let $L_0 \in \mathcal{B}(X, Z)$ be a bounded linear operator such that the mapping $L_0 + \beta E_X^Y : X \to Z$ is bijective for some $\beta \in \mathcal{B}(Y, Z)$.

We are looking for solutions $u \in X$ of problem

$$L_0[u] = \mathcal{F}(u). \tag{8}$$

Let us investigate first in Lemma 2.1 a consequence of the last assumption.

2.1 Lemma *Let* $L_0 + \beta E_X^Y : X \to Z$ *be bijective with some* $\beta \in \mathcal{B}(Y, Z)$. *Then the following implication holds for every* $\gamma \in \mathcal{B}(Y, Z)$:

If $L_0 + \gamma E_X^Y$ *is injective, then it is also surjective, and*

$$(L_0 + \gamma E_X^Y)^{-1} \in \mathcal{B}(Z, X).$$

Proof: Let us assume that $L_0 + \gamma E_X^Y$ is injective. Let $z \in Z$ be given arbitrary. We show that there exists $u \in X$, such that

$$(L_0 + \gamma E_X^Y)\,[u] = z. \tag{9}$$

Equation (9) is equivalent to

$$(L_0 + \beta E_X^Y)\,[u] = (\beta E_X^Y - \gamma E_X^Y)\,[u] + z$$

i.e. to

$$u = Ku + s, \tag{10}$$

with

$$K = (L_0 + \beta E_X^Y)^{-1}\,(\beta E_X^Y - \gamma E_X^Y)\,, \qquad s = (L_0 + \beta E_X^Y)^{-1}\,z.$$

From the Open Mapping Theorem follows that $(L_0 + \beta E_X^Y)^{-1} : Z \to X$ is bounded. Therefore due to the compactness of E_X^Y the operator $K : X \to X$ is also compact. Since the operator $L_0 + \gamma E_X^Y$ is injective, equation (9) has in the homogeneous case (i.e., if $z = 0$) only the trivial solution $u = 0$. Therefore Fredholm's Alternative Theorem yields a unique solution u of (10) for all $z \in Z$. Thus $L_0 + \gamma E_X^Y$ is surjective. Again from the Open Mapping Theorem follows, that $(L_0 + \gamma E_X^Y)^{-1}$ is bounded. □

The following main theorem describes the enclosure method of Plum.

2.2 Theorem (Existence and Enclosure Theorem of Plum) *Let* $\omega \in$

X denote an approximate solution to problem (8) and let us denote the linearisation of the equation at ω by $L : X \to Z$, defined via

$$L = L_0 - \mathcal{F}'(\omega)E_X^Y.$$

Moreover, let the constants δ, C, K and a monotonically nondecreasing function $G : [0,\infty) \to [0,\infty)$ satisfy

$$\|L_0[\omega] - \mathcal{F}(\omega)\|_Z \leq \delta, \tag{11}$$

$$\|u\|_Y \leq C\|u\|_X \qquad \text{for all } u \in X, \tag{12}$$

$$\|u\|_X \leq K\|L[u]\|_Z \qquad \text{for all } u \in X, \tag{13}$$

$$\|\mathcal{F}(\omega + u) - \mathcal{F}(\omega) - \mathcal{F}'(\omega)[u]\|_Z \leq G(\|u\|_Y) \qquad \text{for all } u \in Y, \tag{14}$$

$$G(t) = o(t) \qquad \text{for } t \to 0+. \tag{15}$$

Then, if

$$\delta \leq \frac{1}{K}\alpha - G(C\alpha) \tag{16}$$

holds for some $0 \leq \alpha$, then there exists a solution $u \in X$ of problem (8) satisfying

$$\|u - \omega\|_X \leq \alpha, \tag{17}$$

and in particular

$$\|u - \omega\|_Y \leq C\alpha. \tag{18}$$

Proof: Let us denote the defect of the approximation ω by

$$d = L_0[\omega] - \mathcal{F}(\omega) \in Z. \tag{19}$$

Let $u \in X$ be a solution of (8). Denoting the error by $v = u - \omega \in X$, (19) is equivalent to

$$-d = L_0[v] + \mathcal{F}(\omega) - \mathcal{F}(\omega + v). \tag{20}$$

Let us define $g : Y \to Z$ as

$$g(v) = \mathcal{F}(\omega + v) - \mathcal{F}(\omega) - \mathcal{F}'(\omega)[v]. \tag{21}$$

From (13) follows, that L is injective. According to Lemma 2.1 L is bijective and L^{-1} is bounded. Thus we can rewrite (20) as a fixed point equation

$$T(v) = L^{-1}\left(g(E_X^Y v) - d\right) = v.$$

The operator $T : X \to X$ is continuous and compact, since L^{-1} is bounded, g is continuous and E_X^Y is compact. To apply Schauder's Fixed Point Theorem we are left to find a closed, convex, bounded set $V \subset X$ such that $T(V) \subset V$. We are looking for V as a ball $\{v \in X \colon \|v\|_X \leq \alpha\}$. Due to (11), (12), (13) and (14) we obtain for each $v \in X$ that

$$\|T(v)\|_X \leq K\|L[T(v)]\|_Z = K\|g(E_X^Y v) - d\|_Z$$

$$\leq K\left(\|g(E_X^Y v)\|_Z + \|d\|_Z\right) \leq K\left(G\left(\|E_X^Y v\|_Y\right) + \delta\right) \leq K\left(G\left(C\,\|v\|_X\right) + \delta\right).$$

Now let $\|v\|_X \leq \alpha$. Then due to the monotonicity of G

$$\|T(v)\|_X \leq K\left(G\left(C\,\alpha\right) + \delta\right).$$

Thus $\|T(v)\|_X \leq \alpha$ is fulfilled if

$$K\left(G\left(C\,\alpha\right) + \delta\right) \leq \alpha,$$

i.e., if

$$\delta \leq \frac{1}{K}\,\alpha - G(C\alpha). \tag{22}$$

Thus (22) yields (17).

Inequality (18) follows trivially from (17) and (12). Thus the proof is complete. □

2.3 Remark *Theorem 2.2 holds also in the case, if $\mathcal{F}$ is continuous and Fréchet differentiable only at the approximation ω.*

3 Enclosure for fourth order nonlinear equations

In the first subsection of this section we show how the general Theorem 2.2 can be applied to fourth order nonlinear equations. As we already described in the Introduction, Theorem 2.2 requires approximation ω to the true solution in the space X, in our case in $H_0^2(\Omega)$. In the finite element context this means C^1-approximations. To make the numerical computations easier, we reduce all the computations concerning ω to computations containing only C^0-approximations, as we will describe in Subsection 3.2. In the further subsections we explain how we can obtain the constants δ, C,K and the function G required in Theorem 2.2.

3.1 Application of the method of Plum to fourth order equations

We are aiming at weak solutions of

$$\Delta^2 u = F(u) \qquad \text{on } \Omega, \tag{23}$$

$$u = \frac{\partial u}{\partial \nu} = 0 \qquad \text{on } \partial\Omega,$$

where $\Omega \in \mathbb{R}^2$ is a bounded Lipschitz domain and $F\colon \mathbb{R} \to \mathbb{R}$ is continuously differentiable. Weakly formulated we are looking for $u \in H_0^2(\Omega)$ such that

$$\int_\Omega \Delta u \Delta \varphi \, dx = \int_\Omega F(u)\, \varphi \, dx \quad \text{for all } \varphi \in H_0^2(\Omega).$$

We apply Theorem 2.2 with the following casting: let the Banach spaces be

$$X = (H_0^2(\Omega), \|\cdot\|_{H_0^2}) \text{ with } \|u\|_{H_0^2} = \|\Delta u\|_{L_2},$$

$$Y = (C(\overline{\Omega}), \|\cdot\|_\infty)$$

and $Z = H^{-2}(\Omega)$ equipped with the usual operator norm. Further, let the operator

$$L_0 : H_0^2(\Omega) \to H^{-2}(\Omega), \qquad L_0[u] = \Delta^2 u$$

be defined via

$$L_0[u](\varphi) = \langle u, \varphi \rangle_{H_0^2} = \int_\Omega \Delta u \Delta \varphi \, dx \quad \text{for } \varphi \in H_0^2(\Omega).$$

Let

$$\mathcal{F} : C(\overline{\Omega}) \to H^{-2}(\Omega)$$

be defined via

$$\mathcal{F}(u)(\varphi) = \int_\Omega F(u)\varphi \, dx \quad \text{for } \varphi \in H_0^2(\Omega).$$

Then the mapping $\mathcal{F}$ is Fréchet differentiable in u_0 with Fréchet-derivative

$$\mathcal{F}'(u_0)(h)(\varphi) = \int_\Omega F'(u_0)h\varphi \, dx \qquad \text{for } h \in C(\overline{\Omega}),\ \varphi \in H_0^2(\Omega). \tag{24}$$

According to the Rellich-Kondrachov Imbedding Theorem the imbedding from $H_0^2(\Omega)$ to $C(\overline{\Omega})$ is compact for $n = 2$.

The operator $L_0 = \Delta^2$ is bounded and due to the Riesz Representation Theorem $L_0 + \beta E_{H_0^2(\Omega)}^{C(\overline{\Omega})}$ is bijective for $\beta = 0$.

Thus the setting defined above fulfils the assumptions of Section 2.2.1. To prove existence and enclosure for problem (23) with the help of Theorem 2.2, we have to find the following quantities (i)-(v):

(i) a numerical approximation $\omega \in H_0^2(\Omega)$ to the true solution,
(ii) a constant δ such that

$$\|\Delta^2\omega - \mathcal{F}(\omega)\|_{H^{-2}} \leq \delta, \tag{25}$$

(iii) a constant C such that

$$\|u\|_\infty \leq C\|u\|_{H_0^2} \qquad \text{for all } u \in H_0^2(\Omega), \tag{26}$$

(iv) a constant K such that

$$\|u\|_{H_0^2} \leq K\|\Delta^2 u - \mathcal{F}'(\omega)[u]\|_{H^{-2}} \qquad \text{for all } u \in H_0^2(\Omega), \tag{27}$$

(v) a non-decreasing function $G\colon [0,\infty) \to [0,\infty)$, such that

$$\|\mathcal{F}(\omega+u) - \mathcal{F}(\omega) - \mathcal{F}'(\omega)[u]\|_{H^{-2}} \leq G(\|u\|_\infty) \quad \text{for all } u \in C(\overline{\Omega}), \tag{28}$$

$$G(t) = o(t) \quad \text{for } t \to 0+\,. \tag{29}$$

The differentiability of $\mathcal{F}$ at ω follows then directly from (28) and (29).

We are going to explain in detail in the next subsections, how we can gain the function ω, the constants δ, C and K, and the function G.

3.2 A numerical approximation $\omega \in H_0^2(\Omega)$ to the true solution

Theorem 2.2 requires approximation ω in the space $H_0^2(\Omega)$. Since we compute our approximations with the help of finite element methods, we need to choose appropriate elements. Continuous elements generally do not satisfy the condition being in the space $H_0^2(\Omega)$. Thus to compute ω in $H_0^2(\Omega)$ we would need at least C^1-elements. Implementing C^1-elements is quite difficult and it is moreover numerically expensive. Thus we would like to use only continuous elements. With continuous elements we can certainly not compute ω in $H_0^2(\Omega)$, only approximations in the lower order Sobolev space $H^1(\Omega)$. In order that these approximations suffice for our problem we reformulate it as a system, such that the solutions of this system are in the space $H_0^1(\Omega) \times H^1(\Omega)$. We compute to this reformulated problem numerical approximations with continuous finite elements. Then we define with the help of these approximations the function ω. However, ω will be not computed, all the calculations needed will be led back to the computed approximations.

3.2.1 Reformulation of the problem

We can rewrite problem (23) as a system of equations as

$$
\begin{aligned}
-\Delta u &= v && \text{on } \Omega, && (30)\\
-\Delta v &= F(u) && \text{on } \Omega, && (31)\\
u &= 0 && \text{on } \partial\Omega, && (32)\\
\frac{\partial u}{\partial \nu} &= 0 && \text{on } \partial\Omega. && (33)
\end{aligned}
$$

The weak formulation of this system reads: find $u \in H_0^1(\Omega)$, $v \in H^1(\Omega)$ such that

$$
\begin{aligned}
\int_\Omega \nabla u \cdot \nabla \varphi \, dx &= \int_\Omega v \cdot \varphi \, dx && \text{for all } \varphi \in H^1(\Omega), && (34)\\
\int_\Omega \nabla v \cdot \nabla \psi \, dx &= \int_\Omega F(u) \cdot \psi \, dx && \text{for all } \psi \in H_0^1(\Omega). && (35)
\end{aligned}
$$

Observe, that v does not have any boundary condition, therefore $v \in H^1(\Omega)$, while $u \in H_0^1(\Omega)$ because of (32). To keep the second boundary condition (33) we use test functions $\varphi \in H^1(\Omega)$ in (34), while $\psi \in H_0^1(\Omega)$ in (35) in order not to enforce any boundary condition for v.

Now, to obtain numerical approximations to the weak solution of the system, one only needs C^0-elements. Thus this is a suitable reformulation of our problem.

3.2.2 Numerical approximations

Denote an exact solution of (23) by u^*. Let us compute with the help of the reformulation (34), (35) numerical approximations to u^*, ∇u^*, $-\Delta u^*$ and $-\nabla\Delta u^*$, i.e.,

$$\begin{aligned} \tilde{u} &\approx u^*, & \tilde{u} &\in H_0^1(\Omega), \\ \tilde{\sigma} &\approx \nabla u^*, & \tilde{\sigma} &\in (H_0^1(\Omega))^2, \\ \tilde{v} &\approx -\Delta u^*, & \tilde{v} &\in H^1(\Omega), \\ \tilde{\rho} &\approx -\nabla\Delta u^*, & \tilde{\rho} &\in H(\mathrm{div}, \Omega). \end{aligned}$$

Now let $\omega \in H_0^2(\Omega)$ be defined (not actually computed) via

$$\begin{aligned} \Delta^2\omega &= \Delta \mathit{div}\, \tilde{\sigma}, \\ \text{i.e., } \langle \Delta\omega, \Delta\varphi \rangle_{L_2} &= \langle \mathit{div}\, \tilde{\sigma}, \Delta\varphi \rangle_{L_2}, \quad \text{for all } \varphi \in H_0^2(\Omega). \end{aligned} \tag{36}$$

Existence and uniqueness of ω as a solution of (36) is ensured by the Riesz Representation Theorem. Furthermore, let $\hat{\sigma} \in (H_0^1(\Omega))^2$ be defined by

$$\hat{\sigma} = \nabla\omega. \tag{37}$$

Observe, that because of the definition

$$\mathit{rot}\, \hat{\sigma} = \mathit{rot}\, \nabla\omega = 0, \tag{38}$$

where $\mathit{rot}\, f = \partial_1 f_2 - \partial_2 f_1$ for $f = (f_1, f_2)$.

3.2.3 Estimates for ω

In inequalities (25), (27) and (28) the approximation ω is involved. Since ω is not computed, we can not simply compute K and δ by numerical means. Therefore, we estimate first certain expressions including ω with the help of the computed auxiliary functions $\widetilde{u}$, $\widetilde{\sigma}$, $\widetilde{v}$ and $\widetilde{\rho}$.

To these estimates the existence and computation of the so-called div-rot constant D is essential. The constant D is a value such that

$$\|div\,\sigma\|_{L_2} \leq D\|rot\,\sigma\|_{L_2} \tag{39}$$

holds for all $\sigma \in (H_0^1(\Omega))^2$ with $\Delta div\,\sigma = 0$. In Section 6.1 we prove the existence of D. We also give a method for obtaining a numerically computable upper bound for it on star-shaped domains in Subsection 6.1.1.

In getting estimates of ω the following inequalities play a central role.

3.1 Lemma *With the div-rot constant D of (39) it holds that*

$$\|div\,(\hat{\sigma} - \widetilde{\sigma})\|_{L_2} \leq D \cdot \|rot\,\widetilde{\sigma}\|_{L_2}, \tag{40}$$

$$\|\hat{\sigma} - \widetilde{\sigma}\|_{L_2} \leq C_{H_0^1 \hookrightarrow L_2} \cdot \sqrt{D^2 + 1} \cdot \|rot\,\widetilde{\sigma}\|_{L_2}, \tag{41}$$

where $C_{H_0^1 \hookrightarrow L_2}$ denotes the imbedding constant from $H_0^1(\Omega)$ to $L_2(\Omega)$.

Proof: To prove inequality (40) we apply (39) (or Lemma 6.2). Observe that due to (36) and (37)

$$\Delta div\,(\widetilde{\sigma} - \hat{\sigma}) \overset{(36)}{=} \Delta^2\omega - \Delta div\,\hat{\sigma} \overset{(37)}{=} 0.$$

Therefore taking (38) into account we obtain

$$\|div\,(\widetilde{\sigma} - \hat{\sigma})\|_{L_2} \overset{(39)}{\leq} \|rot\,(\widetilde{\sigma} - \hat{\sigma})\|_{L_2} = \|rot\,\widetilde{\sigma}\|_{L_2}.$$

For the proof of (41) we will need the following representation of the inner product

$$\langle u, v\rangle_{H_0^1} = \langle u_1, v_1\rangle_{H_0^1} + \langle u_2, v_2\rangle_{H_0^1},$$

on $(H_0^1(\Omega))^2$, where $u = (u_1, u_2)$, $v = (v_1, v_2) \in (H_0^1(\Omega))^2$. Then it holds that

$$\langle u, v\rangle_{H_0^1} = \langle \text{div } u, \text{div } v\rangle_{L_2} + \langle \text{rot } u, \text{rot } v\rangle_{L_2} \qquad \text{for } u, v \in (H_0^1(\Omega))^2. \quad (42)$$

For more details we refer to (85). Then using Friedrichs' inequality for $\sigma = (\sigma_1, \sigma_2) \in (H_0^1(\Omega))^2$ we get

$$\begin{aligned} \|\sigma\|_{L_2}^2 = \|\sigma_1\|_{L_2}^2 + \|\sigma_2\|_{L_2}^2 \le C^2_{H_0^1 \hookrightarrow L_2} \left(\|\nabla\sigma_1\|_{L_2}^2 + \|\nabla\sigma_2\|_{L_2}^2\right) = C^2_{H_0^1 \hookrightarrow L_2} \|\sigma\|_{H_0^1}^2 \\ \overset{(42)}{=} C^2_{H_0^1 \hookrightarrow L_2} \left(\|\text{div } \sigma\|_{L_2}^2 + \|\text{rot } \sigma\|_{L_2}^2\right). \end{aligned}$$

With $\sigma = \hat{\sigma} - \widetilde{\sigma}$ in the above calculations we obtain

$$\begin{aligned} \|\widetilde{\sigma} - \hat{\sigma}\|_{L_2}^2 &\le C^2_{H_0^1 \hookrightarrow L_2} \left(\|\text{div } (\widetilde{\sigma} - \hat{\sigma})\|_{L_2}^2 + \|\text{rot } (\widetilde{\sigma} - \hat{\sigma})\|_{L_2}^2\right) \\ &\overset{(40),(38)}{\le} C^2_{H_0^1 \hookrightarrow L_2} \left(D^2 \|\text{rot } \widetilde{\sigma}\|_{L_2}^2 + \|\text{rot } \widetilde{\sigma}\|_{L_2}^2\right) \\ &= C^2_{H_0^1 \hookrightarrow L_2} \left(D^2 + 1\right) \|\text{rot } \widetilde{\sigma}\|_{L_2}^2 . \end{aligned}$$

Thus the proof is complete. □

With the help of the previous lemma we are able now to prove estimates for ω.

3.2 Lemma *For ω defined by (36) the following estimates hold*

$$\begin{aligned} \|\nabla\omega - \widetilde{\sigma}\|_{L_2} &\le C_{H_0^1 \hookrightarrow L_2} \cdot \sqrt{D^2 + 1} \cdot \|\text{rot } \widetilde{\sigma}\|_{L_2}, & (43) \\ \|\nabla\omega - \nabla\widetilde{u}\|_{L_2} &\le \|\widetilde{\sigma} - \nabla\widetilde{u}\|_{L_2} + C_{H_0^1 \hookrightarrow L_2} \cdot \sqrt{D^2 + 1} \cdot \|\text{rot } \widetilde{\sigma}\|_{L_2}, & (44) \\ \|\omega - \widetilde{u}\|_{L_2} &\le C_{H_0^1 \hookrightarrow L_2} \Big(C_{H_0^1 \hookrightarrow L_2} \sqrt{D^2 + 1} \cdot \|\text{rot } \widetilde{\sigma}\|_{L_2} + \|\widetilde{\sigma} - \nabla\widetilde{u}\|_{L_2}\Big) & (45) \end{aligned}$$

Proof: We can prove the above inequalities by the following simple calculations

Inequality (43):

$$\|\nabla\omega - \widetilde{\sigma}\|_{L_2} = \|\hat{\sigma} - \widetilde{\sigma}\|_{L_2} \overset{(41)}{\leq} C_{H_0^1 \hookrightarrow L_2} \cdot \sqrt{D^2+1} \cdot \|rot\ \widetilde{\sigma}\|_{L_2}.$$

Inequality (44):

$$\begin{aligned}\|\nabla\omega - \nabla\widetilde{u}\|_{L_2} &\leq \|\nabla\omega - \widetilde{\sigma}\|_{L_2} + \|\widetilde{\sigma} - \nabla\widetilde{u}\|_{L_2} \\ &\overset{(43)}{\leq} \|\widetilde{\sigma} - \nabla\widetilde{u}\|_{L_2} + C_{H_0^1 \hookrightarrow L_2} \cdot \sqrt{D^2+1} \cdot \|rot\ \widetilde{\sigma}\|_{L_2}.\end{aligned}$$

Inequality (45):

$$\begin{aligned}\|\omega - \widetilde{u}\|_{L_2} &\leq C_{H_0^1 \hookrightarrow L_2} \cdot \|\nabla\omega - \nabla\widetilde{u}\|_{L_2} \\ &\overset{(44)}{\leq} C_{H_0^1 \hookrightarrow L_2} \cdot \left(C_{H_0^1 \hookrightarrow L_2} \cdot \sqrt{D^2+1} \cdot \|rot\ \widetilde{\sigma}\|_{L_2} + \|\widetilde{\sigma} - \nabla\widetilde{u}\|_{L_2}\right).\end{aligned}$$

□

On the right-hand side of the above estimates only computable terms can be found: integrals of approximate functions, the div-rot constant D and an imbedding constant. As already mentioned, we will discuss the computation of the div-rot constant in Section 6.1.1. The approximations $\widetilde{u}$, $\widetilde{\sigma}$, $\widetilde{v}$ and $\widetilde{\rho}$ are numerically computed finite element functions. To calculate the L_2-norm of expressions of these approximations, one needs a validated cubature formula, unless the integrals can be computed in closed form. We give an appropriate formula in Section 6.4.

An upper bound for the imbedding constant $C_{H_0^1 \hookrightarrow L_2}$ can be obtained with the help of a lower bound for the first eigenvalue of the Laplace operator. We are going to deal with this question in Section 6.2. This eigenvalue problem is in general not directly solvable. However, there exist methods to enclose eigenvalues, i.e., to give them an upper and a lower bound. We are going to treat the problem of obtaining these bounds in Section 6.3.

3.3 Computation of δ

Our aim in this section is to find a computable upper bound δ for the defect of ω, i.e.,

$$\|\Delta^2\omega - \mathcal{F}(\omega)\|_{H^{-2}} \leq \delta.$$

Computable means, we are looking for δ in terms of computable quantities, i.e., in terms of $\widetilde{u}$, $\widetilde{v}$, $\widetilde{\sigma}$ and $\widetilde{\rho}$, as well as imbedding constants and the div-rot constant D.

Recall that for $g \in (L_2(\Omega))^2$ and $f \in L_2(\Omega)$ the following equations hold:

$$\|div\ g\|_{H^{-1}} = \sup_{\varphi \in H_0^1(\Omega), \varphi \neq 0} \frac{|\int_\Omega g \nabla\varphi\ dx|}{\|\nabla\varphi\|_{L_2}} \leq \|g\|_{L_2}, \tag{46}$$

and analogously

$$\|\Delta f\|_{H^{-2}} = \sup_{\varphi \in H_0^2(\Omega), \varphi \neq 0} \frac{|\int_\Omega f \Delta\varphi\ dx|}{\|\Delta\varphi\|_{L_2}} \leq \|f\|_{L_2}. \tag{47}$$

3.3 Lemma *Let a constant $C_F > 0$ (depending on $\widetilde{u}$ and ω) be given which satisfies*

$$\|\mathcal{F}(\widetilde{u}) - \mathcal{F}(\omega)\|_{H^{-2}} \leq C_F \|\nabla\widetilde{u} - \nabla\omega\|_{L_2}\ . \tag{48}$$

Then for the defect of ω the following estimate holds:

$$\|\Delta^2\omega - \mathcal{F}(\omega)\|_{H^{-2}} \leq \|div\ \widetilde{\sigma} + \widetilde{v}\|_{L_2} + C_{H_0^1 \hookrightarrow L_2} \cdot \|\nabla\widetilde{v} - \widetilde{\rho}\|_{L_2}$$
$$+ C_{H_0^2 \hookrightarrow L_2} \cdot \|div\ \widetilde{\rho} + F(\widetilde{u})\|_{L_2} + C_F\Big(\|\widetilde{\sigma} - \nabla\widetilde{u}\|_{L_2} + C_{H_0^1 \hookrightarrow L_2} \cdot \sqrt{D^2+1} \cdot \|rot\ \widetilde{\sigma}\|_{L_2} .\Big) \tag{49}$$

Proof: Using the definition of ω, (44) and inequalities (46) and (47), one obtains

$$\begin{aligned}
\|\Delta^2\omega - \mathcal{F}(\omega)\|_{H^{-2}} &= \|\Delta div\,\widetilde{\sigma} - \mathcal{F}(\omega)\|_{H^{-2}} \\
&\leq \|\Delta div\,\widetilde{\sigma} + \Delta\widetilde{v}\|_{H^{-2}} + \|\Delta\widetilde{v} - div\,\widetilde{\rho}\|_{H^{-2}} + \\
&\quad + \|div\,\widetilde{\rho} + \mathcal{F}(\widetilde{u})\|_{H^{-2}} + \|\mathcal{F}(\widetilde{u}) - \mathcal{F}(\omega)\|_{H^{-2}} \\
&\leq \|div\,\widetilde{\sigma} + \widetilde{v}\|_{L_2} + C_{H^{-1}\hookrightarrow H^{-2}} \cdot \|\Delta\widetilde{v} - div\,\widetilde{\rho}\|_{H^{-1}} \\
&\quad + C_{L_2\hookrightarrow H^{-2}} \cdot \|div\,\widetilde{\rho} + F(\widetilde{u})\|_{L_2} + C_F\|\nabla\widetilde{u} - \nabla\omega\|_{L_2} \\
&\leq \|div\,\widetilde{\sigma} + \widetilde{v}\|_{L_2} + C_{H_0^1\hookrightarrow L_2} \cdot \|\nabla\widetilde{v} - \widetilde{\rho}\|_{L_2} + \\
&\quad + C_{H_0^2\hookrightarrow L_2} \cdot \|div\,\widetilde{\rho} + F(\widetilde{u})\|_{L_2} + \\
&\quad + C_F\left(\|\widetilde{\sigma} - \nabla\widetilde{u}\|_{L_2} + C_{H_0^1\hookrightarrow L_2} \cdot \sqrt{D^2+1} \cdot \|rot\,\tilde{\sigma}\|_{L_2}\right).
\end{aligned}$$

In the last step we used the estimates $C_{H^{-1}\hookrightarrow H^{-2}} \leq C_{H_0^1\hookrightarrow L_2}$ and $C_{L_2\hookrightarrow H^{-2}} \leq C_{H_0^2\hookrightarrow L_2}$. For more detail we refer to Section 6.2. □

The right-hand side of (49) contains only numerically computable terms. Moreover the integral expressions are expected to be small. Thus it is an appropriate upper bound δ:

$$\begin{aligned}
&\delta = \|div\,\widetilde{\sigma} + \widetilde{v}\|_{L_2} + C_{H_0^1\hookrightarrow L_2} \cdot \|\nabla\widetilde{v} - \widetilde{\rho}\|_{L_2} \\
&+C_{H_0^2\hookrightarrow L_2} \cdot \|div\,\widetilde{\rho} + F(\widetilde{u})\|_{L_2} + C_F\left(\|\widetilde{\sigma} - \nabla\widetilde{u}\|_{L_2} + C_{H_0^1\hookrightarrow L_2} \cdot \sqrt{D^2+1} \cdot \|rot\,\tilde{\sigma}\|_{L_2}\right).
\end{aligned}$$

3.4 Computation of C

We are looking for a computable upper bound for the imbedding constant from $H_0^2(\Omega)$ to $C(\overline{\Omega})$, i.e., for C such that

$$\|u\|_\infty \leq C\|u\|_{H_0^2} = C\|\Delta u\|_{L_2} \quad \text{for all } u \in H_0^2(\Omega).$$

For this purpose we are going to use a theorem of Plum about L_∞-bounds of functions in the space $H^1(\Omega)$. We assume on the domain $\Omega \subset \mathbb{R}^2$ that it is bounded and that a convex and compact set $Q \subset \mathbb{R}^2$ with $Q^\circ \neq \emptyset$ exists with the following property: for every $x \in \Omega$ there exists an orthogonal matrix $T_x \in \mathbb{R}^{2,2}$ and a vector $b_x \in \mathbb{R}^2$ with

$$x \in T_x Q + b_x \subset \overline{\Omega}\,. \tag{50}$$

Note, that a bounded Lipschitz domain, which we deal with, always satisfies the above condition. Moreover, this property is equivalent to the interior cone condition.

Let us define the moments of a set Q by

$$M_s(Q) = \max_{x_0 \in Q} \left[\frac{1}{|Q|} \int_Q |x - x_0|^s \, dx \right]^{\frac{1}{s}}$$

for $s > 0$. It is easy to see, that $\operatorname{diam}(Q)$ gives a rough upper bound for M_s. Although, if e.g. $s = 1, 2, 4$ and Q is a "nice" domain, e.g. a circular disc or a square, then the integral can be calculated in closed form.

Further, let $p > 2$ and $\alpha > 1.5$ be fixed and define

$$\gamma_0(\alpha) = \frac{\alpha}{2}\sqrt{\frac{\alpha-1}{2\alpha-3}}, \quad \gamma_1(\alpha) = \sqrt{\frac{\alpha}{4\alpha-2}}, \quad \gamma_2(\alpha) = \frac{1}{2}\sqrt{\frac{3}{(2\alpha+1)(\alpha+1)}},$$

$$\begin{aligned}\gamma_3(\alpha, p) &= \frac{1}{2}\left[\frac{\alpha p(3p-2)}{(p-2)(p-1)} \int_0^1 t^{\frac{p-2}{p-1}} (1-t)^{\frac{(\alpha-1)p+1}{p-1}}\right]^{1-\frac{1}{p}} \\ &= \frac{1}{2}\left[\frac{3p-2}{p-2}\frac{\alpha p}{p-1} B\left(\frac{2p-3}{p-1}, \frac{\alpha p}{p-1}\right)\right]^{1-\frac{1}{p}},\end{aligned}$$

where B denotes the Beta-function.

Let us denote by $D[\sigma]$ the Jacobian of $\sigma \in (H^1(\Omega))^2$, and $D_{\text{sym}}[\sigma] = \frac{1}{2}(D[\sigma] + D[\sigma]^T)$. Moreover, let $\|A\|_{F,L_2}$ denote the L_2-Frobenius norm of the function

valued matrix A, i.e., if $A = (a_{ij})_{i,j=1,\ldots,n}$, then $\|A\|_{F,L_2} = \left(\sum_{i,j=1}^{n} \|a_{ij}\|_{L_2}^2\right)^{\frac{1}{2}}$.
The following theorem of Plum gives an upper bound for the L_∞-norm of functions in the space $H^1(\Omega)$, see [24].

3.4 Theorem (Plum) *Let $p > 2$, $u \in H^1(\Omega)$ and $\sigma \in (H^1(\Omega))^2$. Then*

$$\|u\|_\infty \leq C_0\|u\|_{L_2} + C_1\left(\|\nabla u\|_{L_2} + \|\sigma\|_{L_2}\right) + C_2\|D_{\text{sym}}[\sigma]\|_{F,L_2} + C_3\|\nabla u - \sigma\|_{L_p},$$

with constants

$$C_0 = \frac{\gamma_0(\alpha)}{\sqrt{|Q|}},\ C_1 = \frac{\gamma_1(\alpha)M_2(Q)}{\sqrt{|Q|}},\ C_2 = \frac{\gamma_2(\alpha)M_4(Q)^2}{\sqrt{|Q|}},\ C_3 = \frac{\gamma_3(\alpha,p)M_{\frac{p}{p-1}}(Q)}{|Q|^{1/p}}.$$

For the proof we refer to [24].

Observe, that (with suitably chosen α and p) all the numbers γ_i and $M_i(Q)$ are computable, or at least one can easily obtain a good upper bound for them. Thus upper bounds for all the constants C_i are computable ($i = 0, 1, 2, 3$).

3.5 Corollary *For $u \in H_0^2(\Omega)$ it holds that*

$$\|u\|_\infty \leq (C_0 \cdot C_{H_0^2 \hookrightarrow L_2} + 2 \cdot C_1 \cdot C_{H_0^2 \hookrightarrow H_0^1} + C_2) \cdot \|\Delta u\|_{L_2} = C \cdot \|\Delta u\|_{L_2}. \quad (51)$$

Proof: Let us choose $\sigma = \nabla u$ in Theorem 3.4. Then we get the estimate

$$\|u\|_\infty \leq C_0\|u\|_{L_2} + 2 \cdot C_1\|\nabla u\|_{L_2} + C_2\|u_{xx}\|_{F,L_2}, \quad (52)$$

where u_{xx} denotes the Hesse-matrix of u.

Moreover, for $u \in H_0^2(\Omega)$ it holds that

$$\|u\|_{L_2} \leq C_{H_0^2 \hookrightarrow L_2} \|\Delta u\|_{L_2}, \tag{53}$$

$$\|\nabla u\|_{L_2} \leq C_{H_0^2 \hookrightarrow H_0^1} \|\Delta u\|_{L_2}, \tag{54}$$

$$\|u_{xx}\|_{F,L_2} = \|\Delta u\|_{L_2}. \tag{55}$$

Only the third inequality has to be proven. Let $u \in H_0^2(\Omega)$ and with $\sigma = (\sigma_1, \sigma_2) = \nabla u \in (H_0^1(\Omega))^2$ we get

$$\|u_{xx}\|_{F,L_2}^2 = \left\|\frac{\partial^2 u}{\partial x^2}\right\|_{L_2}^2 + \left\|\frac{\partial^2 u}{\partial x \partial y}\right\|_{L_2}^2 + \left\|\frac{\partial^2 u}{\partial y \partial x}\right\|_{L_2}^2 + \left\|\frac{\partial^2 u}{\partial y^2}\right\|_{L_2}^2 = \|\nabla \sigma_1\|_{L_2}^2 + \|\nabla \sigma_2\|_{L_2}^2$$

$$= \|\sigma\|_{H_0^1}^2 \overset{(85)}{=} \|div\ \sigma\|_{L_2}^2 + \|rot\ \sigma\|_{L_2}^2 = \|div\ \sigma\|_{L_2}^2 = \|\Delta u\|_{L_2}^2.$$

If we combine (52) with (53), (54) and (55), we get

$$\|u\|_\infty \leq (C_0 \cdot C_{H_0^2 \hookrightarrow L_2} + 2 \cdot C_1 \cdot C_{H_0^2 \hookrightarrow H_0^1} + C_2) \cdot \|\Delta u\|_{L_2}. \tag{56}$$

□

Let us estimate now the constants C_0, C_1, C_2 and C_3 occurring in Theorem 3.4. (We do not need the constant C_3 for the computation of C, but we will need it later in Section 4.3.)

If we analyse the functions $\gamma_0(\alpha)$, $\gamma_1(\alpha)$, $\gamma_2(\alpha)$ and $\gamma_3(\alpha, p)$, we discover that on $]\frac{3}{2}, \infty[$ the function γ_0 takes its minimum in $\alpha = 2$ and it is monoton increasing on $[2, \infty[$, the function γ_1 is monotone decreasing on $]\frac{1}{2}, \infty[$ and the function γ_2 is monotone decreasing on $]-\frac{1}{2}, \infty[$. For γ_3 observe, that the function

$$\frac{1}{2}\left[\frac{3p-2}{p-2}\frac{\alpha p}{p-1}\frac{1}{\alpha(\alpha+1)}\right]^{1-\frac{1}{p}}$$

is monotone decreasing in p on $]2, \infty[$ and also in α on $]0, \infty[$.

Thus $p = 4$ and $\alpha = 3$ is a possible good choice. With these values one obtains

$$\gamma_0 = \sqrt{\frac{3}{2}}, \qquad \gamma_1 = \sqrt{\frac{3}{10}}, \qquad \gamma_2 = \frac{1}{4}\sqrt{\frac{3}{7}},$$

$$\gamma_3 = \frac{1}{2} \cdot \left(20 \int_0^1 t^{\frac{2}{3}}(1-t)^3\right)^{\frac{3}{4}} = \frac{27}{2}\left(\frac{3}{154}\right)^{\frac{3}{4}}.$$

Let us calculate now the moments M_2, M_4 and M_s in case of a circular disc of radius ρ and of a square with side length a. It is easy to see, that the function $f(y) = \int_Q |x-y|^\lambda \, dx$ is subharmonic, since

$$\Delta f(y) = \int_Q \Delta_y |x-y|^\lambda \, dx = \int_Q \mathrm{div}_y \left(\lambda |x-y|^{\lambda-2}(y-x)\right) dx =$$

$$= \int_Q \lambda^2 |x-y|^{\lambda-2} \, dx > 0$$

Thus f takes its maximum on the boundary of Q. In case of the circular disc with radius ρ, using its symmetry and transforming to the unit disc E_0, its moments can be reduced to

$$M_s(Q) = \left[\frac{1}{|Q|}\int_Q |x-(\rho,0)|^s \, dx\right]^{\frac{1}{s}} = \rho M_s(E_\circ).$$

Analogously in case of the square with side length a we obtain

$$M_s(Q) = \left[\frac{1}{|Q|}\int_Q |x|^s \, dx\right]^{\frac{1}{s}} = a M_s(E_\square),$$

where $E_\square$ denotes the unit square.

The moments M_2 and M_4 can be calculated in closed form, but in general the term $M_{\frac{p}{p-1}}$ do not. One can apply a quadrature formula with remainder term to obtain an upper bound for its value. We show another possible formula in Section 6.4.1.

It follows then for $p = 4$ that

$$M_2(Q) = \sqrt{\frac{3}{2}}\rho \qquad M_4(Q) = \sqrt[4]{\frac{10}{3}}\rho \quad \text{and} \quad M_{4/3}(Q) \leq \rho \cdot 1.1672 \tag{57}$$

in case of a circular disc with radius ρ, and

$$M_2(Q) = \sqrt{\frac{2}{3}}a \qquad M_4(Q) = \sqrt[4]{\frac{28}{45}}a \quad \text{and} \quad M_{4/3}(Q) \leq a \cdot 0.8497 \tag{58}$$

in case of a square with side length a.

Now we can calculate the desired constants in case of a circular disc with radius ρ:

$$C_0 = \sqrt{\frac{3}{2\pi}\frac{1}{\rho}},\ C_1 = \frac{3}{2\sqrt{5\pi}},\ C_2 = \frac{1}{4}\sqrt{\frac{10}{7\pi}}\rho,\ C_3 = \frac{27}{2}\left(\frac{3}{154}\right)^{\frac{3}{4}}\frac{\sqrt{\rho}}{\pi^{1/4}}\,1.1672. \tag{59}$$

Then according to (51) we have

$$C(\rho) = \sqrt{\frac{3}{2\pi}} \cdot C_{H_0^2 \hookrightarrow L_2}\frac{1}{\rho} + 2 \cdot \frac{3}{2\sqrt{5\pi}} \cdot C_{H_0^2 \hookrightarrow H_0^1} + \frac{1}{4}\sqrt{\frac{10}{7\pi}}\rho.$$

The optimal radius ρ that minimises $C(\rho)$ can be determined for each domain Ω. The minimum of $C(\rho)$ on $[0, \infty)$ will be attained in

$$\rho_{min} = \sqrt{2\sqrt{\frac{21}{5}}C_{H_0^2 \hookrightarrow L_2}}\,. \tag{60}$$

Furthermore, an upper bound for the radius arises from condition (50). One chooses for ρ the smaller of these two values.

In case of a square we have

$$C_0 = \sqrt{\frac{3}{2}\frac{1}{a}},\quad C_1 = \frac{1}{\sqrt{5}},\quad C_2 = \frac{1}{2\sqrt{15}}a,\quad C_3 = \frac{27}{2}\left(\frac{3}{154}\right)^{\frac{3}{4}} 0.8497\sqrt{a},$$

and thus

$$C(a) = \sqrt{\frac{3}{2}} \cdot C_{H_0^2 \hookrightarrow L_2} \frac{1}{a} + \frac{2}{\sqrt{5}} \cdot C_{H_0^2 \hookrightarrow H_0^1} + \frac{1}{2\sqrt{15}} a.$$

One can choose for ρ the smaller value of the minimiser of $C(a)$ on $[0, \infty)$ that is

$$\sqrt{3\sqrt{10}\, C_{H_0^2 \hookrightarrow L_2}}, \tag{61}$$

and the upper bound arising from condition (50).

3.5 Computation of K

Let us introduce a new scalar product in $H_0^2(\Omega)$, namely for $\alpha > 0$ let

$$\langle u, v \rangle_{H_0^2,\alpha} = \int_\Omega \Delta u \Delta v + \alpha u v \, dx \qquad \text{for } u, v \in H_0^2(\Omega).$$

Let us denote the corresponding norm in $H_0^2(\Omega)$ by $\| \cdot \|_{H_0^2,\alpha}$ and the corresponding operator norm in $H^{-2}(\Omega)$ by $\| \cdot \|_{H^{-2},\alpha}$. We will make clear the role of α in Section 6.3.5.

We are looking now for a constant K satisfying the inequality

$$\|u\|_{H_0^2,\alpha} \leq K \|L[u]\|_{H^{-2},\alpha} \qquad \text{for all } u \in H_0^2(\Omega), \tag{62}$$

where L is the linearisation of our problem at ω:

$$Lu = \Delta^2 u - F'(\omega) u = \Delta^2 u - c u,$$

with $c = F'(\omega)$. Due to

$$\|u\|_{H_0^2} \leq \|u\|_{H_0^2,\alpha} \leq K \|L[u]\|_{H^{-2},\alpha} = K \sup_{\varphi \in H_0^2(\Omega) \setminus \{0\}} \frac{|L[u](\varphi)|}{\|\varphi\|_{H_0^2,\alpha}}$$

$$\leq K \sup_{\varphi \in H_0^2(\Omega)\setminus\{0\}} \frac{|L[u](\varphi)|}{\|\varphi\|_{H_0^2}} = K\|L[u]\|_{H^{-2}},$$

this K satisfies (27).

The operator L depends on the function ω, which has not been computed, only defined via (36). Thus it would be difficult to construct K directly from inequality (62). Instead, we define similarly to the operator L the perturbed operator $\widetilde{L}\colon H_0^2(\Omega) \to H^{-2}(\Omega)$ with $\widetilde{u}$ in place of ω by

$$\widetilde{L}[u] \;=\; \Delta^2 u - F'(\widetilde{u})u = \Delta^2 u - \widetilde{c}u,$$

denoting analogously $F'(\widetilde{u})$ by $\widetilde{c}$.

Then we calculate first a constant $\widetilde{K}$ similar to K and a constant γ such that for all $u \in H_0^2(\Omega)$ hold

$$\|u\|_{H_0^2,\alpha} \;\leq\; \widetilde{K}\|\widetilde{L}[u]\|_{H^{-2},\alpha}, \tag{63}$$

$$\|(\widetilde{L} - L)[u]\|_{H^{-2},\alpha} \;\leq\; \gamma\|u\|_{H_0^2,\alpha}. \tag{64}$$

Then we obtain for all $u \in H_0^2(\Omega)$ that

$$\begin{aligned}\|u\|_{H_0^2,\alpha} \leq \widetilde{K}\|\widetilde{L}[u]\|_{H^{-2},\alpha} &\leq \widetilde{K}\big(\|L[u]\|_{H^{-2},\alpha} + \|(\widetilde{L} - L)[u]\|_{H^{-2},\alpha}\big)\\ &\leq \widetilde{K}\|L[u]\|_{H^{-2},\alpha} + \widetilde{K}\gamma\|u\|_{H_0^2,\alpha}.\end{aligned}$$

This yields in case $\gamma\widetilde{K} < 1$

$$\|u\|_{H_0^2,\alpha} \leq \underbrace{\frac{\widetilde{K}}{1-\gamma\widetilde{K}}}_{=:K}\|L[u]\|_{H^{-2},\alpha}.$$

Observe, that if $\widetilde{u}$ is a good approximation of ω, then $\widetilde{L}$ is "close" to L. Therefore γ can be chosen small enough to fulfil $\gamma\widetilde{K} < 1$. We will show in

Section 4 a possible way to find an upper bound for γ in case of the Gelfand-equation, and in Section 5 in case of the Emden-equation.

Let us consider now the problem of finding the (optimal) constant $\widetilde{K}$ in (63). This problem is equivalent to finding a positive lower bound for

$$\min_{u\in H_0^2(\Omega),u\neq 0} \frac{\|\widetilde{L}[u]\|_{H^{-2},\alpha}^2}{\|u\|_{H_0^2,\alpha}^2}. \tag{65}$$

To abolish the operator norm let us introduce the canonical isomorphism $\Phi\colon H_0^2(\Omega)\to H^{-2}(\Omega)$ with respect to the norm $\|\cdot\|_{H_0^2,\alpha}$, i.e.,

$$\Phi(\varphi)(\psi) = (\Delta^2\varphi + \alpha\varphi)[\psi] = \langle\varphi,\psi\rangle_{H_0^2,\alpha}.$$

Then $\|\widetilde{L}[u]\|_{H^{-2},\alpha} = \|\Phi^{-1}\widetilde{L}[u]\|_{H_0^2,\alpha}$, therefore problem (65) becomes to find a positive lower bound for

$$\min_{u\in H_0^2(\Omega),u\neq 0} \frac{\|\Phi^{-1}\widetilde{L}[u]\|_{H_0^2,\alpha}^2}{\|u\|_{H_0^2,\alpha}^2}.$$

Since

$$\langle\Phi^{-1}\widetilde{L}[u],v\rangle_{H_0^2,\alpha} = (\widetilde{L}[u])[v] = \int_\Omega \Delta u\Delta v - \widetilde{c}uv\,dx = \langle u,\Phi^{-1}\widetilde{L}v\rangle_{H_0^2,\alpha},$$

the operator $\Phi^{-1}\widetilde{L}$ is symmetric. As it is defined on the whole space $H_0^2(\Omega)$, it is self-adjoint. Therefore (62) holds with

$$K \geq \left[\min\left\{|\mu|\colon \mu\in\sigma\left(\Phi^{-1}\widetilde{L}\right)\right\}\right]^{-1},$$

where $\sigma\left(\Phi^{-1}\widetilde{L}\right)$ denotes the spectrum of the operator $\Phi^{-1}\widetilde{L}$. Let us consider therefore the eigenvalue problem

$$\Phi^{-1}\widetilde{L}[u] = \mu u, \qquad u\in H_0^2(\Omega). \tag{66}$$

This problem is equivalent to

$$\Delta^2 u - \widetilde{c}u = \mu(\Delta^2 u + \alpha u), \qquad u \in H_0^2(\Omega),$$

i.e. to

$$(\alpha + \widetilde{c})u = (1 - \mu)(\Delta^2 u + \alpha u), \qquad u \in H_0^2(\Omega).$$

Thus we obtained the new eigenvalue problem

$$u \in H_0^2(\Omega), \int_\Omega (\widetilde{c}(x) + \alpha)u\varphi \, dx = \kappa \int_\Omega \Delta u \Delta\varphi + \alpha u\varphi \, dx, \quad \text{for all } \varphi \in H_0^2(\Omega), \tag{67}$$

with $\kappa = 1 - \mu$.

In order to ensure that all the eigenvalues of (67) are positive, we assume from now on that α is chosen such that

$$\widetilde{c}(x) + \alpha > 0 \qquad \text{for all } x \in \overline{\Omega}. \tag{68}$$

Analogously to the corresponding eigenvalue problem in [21] for equations of second order, we can state the following lemma:

3.6 Lemma *The spectrum of problem (67) consists of the point zero and a countable number of positive real eigenvalues accumulating only at zero.*

Proof: Choose any $p > 1$ and let $q = \frac{2p}{p-1}$. The imbedding $H_0^2(\Omega) \hookrightarrow L_q(\Omega)$ is compact, therefore for $u \in L_q(\Omega)$, $\varphi \in H_0^2(\Omega)$ there exists a constant d such that

$$\left| \int_\Omega (\widetilde{c} + \alpha)u\varphi \, dx \right| \leq ||\widetilde{c} + \alpha||_p ||u||_q ||\varphi||_q \leq d||u||_q ||\varphi||_{H_0^2,\alpha}.$$

Thus the functional $\varphi \to \int_\Omega (\widetilde{c} + \alpha)u\varphi \, dx$ is linear and bounded on $H_0^2(\Omega)$.

We can apply then Lax-Milgram Lemma for the problem

$$\psi \in H_0^2(\Omega), \qquad \int_\Omega \Delta\psi\Delta\varphi + \alpha\psi\varphi \, dx = \int_\Omega (\widetilde{c} + \alpha)u\varphi \, dx \quad \text{for all } \varphi \in H_0^2(\Omega). \tag{69}$$

Let us denote by Tu the unique solution ψ of (69) for every $u \in L_q$. T is then a linear operator from L_q to $H_0^2(\Omega)$. Inserting $\varphi = Tu$ into (69) we get

$$\|Tu\|^2_{H_0^2,\alpha} = \int_\Omega \Delta Tu\Delta Tu + \alpha TuTu \, dx = \int_\Omega (\widetilde{c} + \alpha)uTu \, dx \leq d\|u\|_q\|Tu\|_{H_0^2,\alpha},$$

i.e., T is bounded. The operator $\widetilde{T} = TE^{L_q}_{H_0^2} \colon H_0^2(\Omega) \to H_0^2(\Omega)$ is compact and since

$$\langle Tu, v\rangle_{H_0^2,\alpha} = \int_\Omega \Delta Tu\Delta v + \alpha(Tu)v \, dx = \int_\Omega (\widetilde{c} + \alpha)uv \, dx = \langle u, Tv\rangle_{H_0^2,\alpha}$$

holds for all $u, v \in H_0^2(\Omega)$, it is also symmetric. Taking (68) into account we can deduce that the spectrum of $\widetilde{T}$ consists of the point zero and a countable number of positive real eigenvalues accumulating at zero. Since the eigenvalue problem $\widetilde{T}u = \kappa u$ is equivalent to (67), we are done. □

Returning to problem (66) we obtain from Lemma 3.6 and from $\mu = 1 - \kappa$ that $\Phi^{-1}\widetilde{L}$ has countably many eigenvalues with accumulation point 1. We have

$$\mu_0 = \min\{|\mu| : \mu \in \sigma(\Phi^{-1}\widetilde{L})\} = \min\{|1 - \kappa| : \kappa \text{ eigenvalue of } (67)\}.$$

To give an upper bound for $\widetilde{K}$ in terms of the above eigenvalues let us denote

$$A_{\text{upper}} := \inf\{\kappa : \kappa \text{ eigenvalue of } (67),\ \kappa > 1\},$$

$$A_{\text{lower}} := \max\{\kappa : \kappa \text{ eigenvalue of } (67),\ \kappa < 1\},$$

with $\inf \emptyset = +\infty$. Then we get the upper bound for $\widetilde{K}$

$$\widetilde{K} \leq \frac{1}{\min\{1 - A_{\text{lower}}, A_{\text{upper}} - 1\}}.$$

As the eigenvalue problem (67) is not directly solvable, we need eigenvalue bounds. We discuss the methods for obtaining such bounds in Section 6.3.

3.6 Determination of the function G

We are looking for a non-decreasing function $G\colon [0,\infty) \to [0,\infty)$, such that

$$\|\mathcal{F}(\omega + u) - \mathcal{F}(\omega) - \mathcal{F}'(\omega)[u]\|_{H^{-2}} \leq G(\|u\|_\infty) \qquad \text{for all } u \in C(\overline{\Omega}), \tag{70}$$

and

$$G(t) = o(t) \quad \text{for } t \to 0+$$

hold. If we replace in (70) the terms of u by $y \in \mathbb{R}$ and $\mathcal{F}$ by F, we get the following similar, but simpler inequality for a function $\widetilde{G}\colon [0,\infty) \to [0,\infty)$:

$$\big|F(\omega(x) + y) - F(\omega(x)) - F'(\omega(x))y\big| \leq \widetilde{G}(|y|) \qquad \text{for all } x \in \Omega,\ y \in \mathbb{R}. \tag{71}$$

Assuming that such a function $\widetilde{G}$ exists, which is in addition non-decreasing and satisfies

$$\widetilde{G}(t) = o(t) \quad \text{for } t \to 0+, \tag{72}$$

then we can obtain the desired function G as follows:

$$\|\mathcal{F}(\omega + u) - \mathcal{F}(\omega) - \mathcal{F}'(\omega)[u]\|_{H^{-2}}$$

$$
\begin{aligned}
&= \sup_{\varphi \in H_0^2(\Omega)\setminus\{0\}} \frac{\left| \int_\Omega \big(F(\omega(x)+u(x)) - F(\omega(x)) - F'(\omega(x))u(x)\big)\varphi(x)\,dx \right|}{\|\Delta\varphi\|_{L_2}} \\
&\overset{(71)}{\leq} \sup_{\varphi \in H_0^2(\Omega)\setminus\{0\}} \frac{\int_\Omega \widetilde{G}(|u(x)|)|\varphi(x)|\,dx}{\|\Delta\varphi\|_{L_2}} \\
&\leq \sup_{\varphi \in H_0^2(\Omega)\setminus\{0\}} \frac{\|\widetilde{G}(|u|)\|_\infty \int_\Omega |\varphi(x)|\,dx}{\|\Delta\varphi\|_{L_2}} \\
&\overset{\widetilde{G}\text{ mon.}}{\leq} |\Omega|^{1/2} \cdot C_{H_0^2 \hookrightarrow L_2} \cdot \widetilde{G}(\|u\|_\infty)
\end{aligned}
$$

for all $u \in C(\overline{\Omega})$. Then define $G(t) = |\Omega|^{1/2} \cdot C_{H_0^2 \hookrightarrow L_2} \cdot \widetilde{G}(t)$.

The advantage of this method is that it is easier to find $\widetilde{G}$ first, as to find G directly.

In Section 4 we will show how we can find a function $\widetilde{G}$ in the case of the Gelfand-equation, and in Section 5 in the case of the Emden-equation.

4 Application to the Gelfand-equation

As an application of the previous results let us consider the fourth order Gelfand-equation

$$
\begin{aligned}
\Delta^2 u &= \lambda\, e^u \quad \text{on } \Omega, \\
u = \frac{\partial u}{\partial \nu} &= 0 \quad \text{on } \partial\Omega,
\end{aligned}
\tag{73}
$$

with $\lambda \geq 0$.

The function $F(u) = \lambda e^u$ clearly fulfils the regularity assumptions we made in Section 3.1. In the following subsections we determine the quantities depending on the function F, namely the constants δ and γ and the function $\widetilde{G}$. Afterwards we demonstrate numerical results on the unit-square, on a disc-like domain and on a dumbbell-like domain, which verify the existence of solutions of the Gelfand-equation.

4.1 Computation of δ

Due to Lemma 3.3 for determining δ we only have to determine the constant C_F satisfying (48). We will make use of the Trudinger-Moser inequality:

4.1 Lemma (Trudinger-Moser inequality) *If $u \in H_0^1(\Omega)$ and $c > 1/\sqrt{4\pi}$ then*

$$\frac{1}{|\Omega|}\int_\Omega \exp\left(\frac{u^2(x)}{c^2\|u\|_{H_0^1}^2}\right) dx \leq 1 + \frac{1}{4\pi c^2 - 1}. \tag{74}$$

For the proof see [16], (Theorem 1 and the first part of its proof). The following Corollary is due to Plum and Wieners, see [26].

4.2 Corollary *If $u \in H_0^1(\Omega)$, then $\exp(|u|) \in L_q(\Omega)$ for $1 < q < \infty$. Moreover, it holds that*

$$\|\exp(|u|)\|_{L_q} \leq |\Omega|^{\frac{1}{q}} \left(1 + \frac{1}{4\pi c^2 - 1}\right)^{\frac{1}{q}} \exp\left(\frac{qc^2\|u\|_{H_0^1}^2}{4}\right),$$

with $c > 1/\sqrt{4\pi}$.

Proof: Let $1 < q < \infty$ and $c > 1/\sqrt{4\pi}$. From the arithmetic-geometric mean inequality follows for $0 \leq t$, $d \neq 0$ that

$$qt \leq q^2 d^2/4 + t^2/d^2,$$

which yields

$$\exp(qt) \leq \exp(q^2 d^2/4) \cdot \exp(t^2/d^2).$$

On substituting $t = |u(x)|$ and $d = c\|u\|_{H_0^1}$ and integrating one gets

$$\int_\Omega \exp(|u(x)|)^q \, dx \leq \exp\left(\frac{q^2c^2\|u\|_{H_0^1}^2}{4}\right) \cdot \int_\Omega \exp\left(\frac{u(x)^2}{c^2\|u\|_{H_0^1}^2}\right) dx$$

$$\overset{(74)}{\leq} |\Omega| \left(1 + \frac{1}{4\pi c^2 - 1}\right) \exp\left(\frac{q^2 c^2 \|u\|^2_{H_0^1}}{4}\right).$$

□

The following lemma is a consequence of Corollary 4.2.

4.3 Lemma *For $u, v \in H_0^1(\Omega)$ it holds that*

$$\|e^u - e^v\|_{L_2} \leq K_1 \cdot C_{H_0^1 \hookrightarrow L_6} \cdot \|e^u\|_{L_6} \cdot \exp\left(1.5 \cdot \|\nabla u - \nabla v\|^2_{L_2}\right) \cdot \|\nabla u - \nabla v\|_{L_2},$$

with

$$K_1 = |\Omega|^{\frac{1}{6}} \left(1 + \frac{1}{4\pi^2 - 1}\right)^{\frac{1}{6}}.$$

Proof: Using Hölder's inequality and the mean value theorem we get

$$\begin{aligned}
\|e^u - e^v\|^2_{L_2} &= \int_\Omega |e^u - e^v|^2 \, dx \\
&= \int_\Omega e^{2u} \left|1 - e^{v-u}\right|^2 \, dx \\
&\overset{\text{MVT}}{\leq} \int_\Omega e^{2u} \cdot e^{2|v-u|} \cdot |v - u|^2 \, dx \\
&\leq \|e^{2u}\|_{L_3} \cdot \|e^{2|v-u|}\|_{L_3} \cdot \left\||v - u|^2\right\|_{L_3} \\
&= \|e^u\|^2_{L_6} \cdot \|e^{|v-u|}\|^2_{L_6} \cdot \|v - u\|^2_{L_6} \\
&\leq \|e^u\|^2_{L_6} \cdot \|e^{|v-u|}\|^2_{L_6} \cdot C^2_{H_0^1 \hookrightarrow L_6} \|\nabla v - \nabla u\|^2_{L_2}.
\end{aligned}$$

We obtain from Corollary 4.2 with $q = 6$ and by choosing $c = 1$ for $\varphi \in H_0^1(\Omega)$ that

$$\|e^{|\varphi|}\|_{L_6} \leq \underbrace{|\Omega|^{\frac{1}{6}} \left(1 + \frac{1}{4\pi^2 - 1}\right)^{\frac{1}{6}}}_{=:K_1} \exp\left(1.5 \cdot \|\varphi\|^2_{H_0^1}\right).$$

Therefore

$$\|e^{|v-u|}\|_{L_6} \leq K_1 \exp\left(1.5 \cdot \|\nabla v - \nabla u\|^2_{L_2}\right).$$

Thus the assertion holds. $\square$

Now we can state our lemma about the constant C_F satisfying (48).

4.4 Lemma *The constant C_F defined as*

$$\begin{aligned} C_F &= |\lambda| \cdot K_1 \cdot C_{H_0^2 \hookrightarrow L_2} \cdot C_{H_0^1 \hookrightarrow L_6} \cdot \|e^{\widetilde{u}}\|_{L_6} \\ &\quad \cdot \exp\left(1.5 \cdot \left(\|\nabla \widetilde{u} - \widetilde{\sigma}\|_{L_2} + C_{H_0^1 \hookrightarrow L_2} \sqrt{D^2+1} \cdot \|\mathrm{rot}\, \widetilde{\sigma}\|_{L_2}\right)^2\right), \end{aligned}$$

with K_1 from Lemma 4.3, fulfils the requirement of (48).

Proof: Using the boundedness of the imbedding from $H_0^2(\Omega)$ to $L_2(\Omega)$ one gets

$$\begin{aligned} \frac{1}{|\lambda|}\|\mathcal{F}(\widetilde{u}) - \mathcal{F}(\omega)\|_{H^{-2}} &= \sup_{\varphi \in H_0^2(\Omega)\setminus\{0\}} \frac{|\int_\Omega (F(\widetilde{u}) - F(\omega))\varphi \, dx|}{\|\Delta\varphi\|_{L_2}} \\ &\le C_{H_0^2 \hookrightarrow L_2} \|F(\widetilde{u}) - F(\omega)\|_{L_2} \le C_{H_0^2 \hookrightarrow L_2} \cdot \|e^{\widetilde{u}} - e^{\omega}\|_{L_2}. \end{aligned}$$

Using Lemma 4.3 and (44) we obtain the assertion. $\square$

4.2 Computation of γ

In case of the Gelfand-equation one can find an appropriate constant γ satisfying (64) as follows.

4.5 Lemma *The constant γ defined as*

$$\gamma = |\lambda| \, C^2_{H_0^2 \hookrightarrow L_4} \|e^{\widetilde{u}} - e^{\omega}\|_{L_2}$$

fulfils the requirement of (64).

Proof: With the following straightforward calculations we get

$$\begin{aligned}
\|(\widetilde{L}-L)[u]\|_{H^{-2},\alpha} &= \|F'(\widetilde{u})[u]-F'(\omega)[u]\|_{H^{-2},\alpha} \\
&= |\lambda|\,\|e^{\widetilde{u}}[u]-e^{\omega}[u]\|_{H^{-2},\alpha} \\
&= |\lambda| \sup_{\varphi\in H_0^2(\Omega),\,\varphi\neq 0} \frac{|\int_\Omega (e^{\widetilde{u}}-e^{\omega})u\varphi\,dx|}{\|\varphi\|_{H_0^2,\alpha}} \\
&= |\lambda| \sup_{\varphi\in H_0^2(\Omega),\,\varphi\neq 0} \frac{\|e^{\widetilde{u}}-e^{\omega}\|_{L_2}\|u\|_{L_4}\|\varphi\|_{L_4}}{\|\varphi\|_{H_0^2,\alpha}} \\
&\le |\lambda|\, C^2_{H^2_{0,\alpha}\hookrightarrow L_4}\|e^{\widetilde{u}}-e^{\omega}\|_{L_2}\|u\|_{H_0^2,\alpha} \\
&\le |\lambda|\, C^2_{H^2_0\hookrightarrow L_4}\|e^{\widetilde{u}}-e^{\omega}\|_{L_2}\|u\|_{H_0^2,\alpha}.
\end{aligned}$$

□

One can compute γ by using Lemma 4.3 for the term $\|e^{\widetilde{u}}-e^{\omega}\|_{L_2}$.

4.3 Determination of the function $\widetilde{G}$

We show how we can determine the function $\widetilde{G}$ fulfilling the requirements of Section 3.6.

4.6 Lemma *The function* $\widetilde{G}: [0,\infty[\to [0,\infty[$, *defined as*

$$\widetilde{G}(y)=\lambda\cdot e^{\|\omega\|_\infty}\cdot(e^y-y-1)$$

is non-decreasing and it fulfils (71) and (72).

Proof: It is easy to see, that the function

$$f(y)=e^y-y-1 \tag{75}$$

is nonnegative, monotone decreasing on $(-\infty,0)$, monotone increasing on $[0,\infty)$, it fulfils $f(y)=o(y)$ as $y\to 0$ and $f(y)\le f(|y|)$ for all $y\in\mathbb{R}$. Using these properties we obtain

$$|F(\omega(x)+y)-F(\omega(x))-F'(\omega(x))y| =$$

$$|\lambda e^{\omega(x)+y}-\lambda e^{\omega(x)}-\lambda e^{\omega(x)}y| = \lambda\cdot e^{\omega(x)}|e^y-y-1| \leq \lambda\cdot e^{\|\omega\|_\infty}(e^y-y-1).$$

Again, from the properties of f follows, that $\widetilde{G}$ fulfils the assumptions we made. □

To finish this section we are left to find a computable upper bound for $\|\omega\|_\infty$.

4.7 Lemma *With constants C_0, C_1, C_2, C_3 from Theorem 3.4 it holds that*

$$\|\omega\|_\infty \leq \|\widetilde{u}\|_\infty + C_0\|\omega-\widetilde{u}\|_{L_2} + C_1(\|\nabla(\omega-\widetilde{u})\|_{L_2} + \|\hat{\sigma}-\widetilde{\sigma}\|_{L_2}) +$$

$$C_2\sqrt{D^2+\frac{1}{2}\|rot\ \widetilde{\sigma}\|_{L_2}} + C_3\|\nabla\widetilde{u}-\widetilde{\sigma}\|_{L_p}. \tag{76}$$

Proof: Due to

$$\|\omega\|_\infty \leq \|\omega-\widetilde{u}\|_\infty + \|\widetilde{u}\|_\infty,$$

it is enough to bound $\|\omega-\widetilde{u}\|_\infty$ from above. We can fulfil this task with the help of Theorem 3.4. Let $u=\omega-\widetilde{u}$ and $\sigma=\hat{\sigma}-\widetilde{\sigma}$ in the notations of the theorem. Then $u\in H_0^1(\Omega)$ and $\sigma\in(H_0^1(\Omega))^2$, thus u and σ satisfy the assumptions of the theorem. Therefore using $\hat{\sigma}=\nabla\omega$ it holds that

$$\|\omega-\widetilde{u}\|_\infty \leq C_0\|\omega-\widetilde{u}\|_{L_2} + C_1(\|\nabla(\omega-\widetilde{u})\|_{L_2} + \|\hat{\sigma}-\widetilde{\sigma}\|_{L_2})$$

$$+ C_2\|D_{\text{sym}}[\hat{\sigma}-\widetilde{\sigma}]\|_{L_2} + C_3\|\nabla\widetilde{u}-\widetilde{\sigma}\|_{L_p}.$$

For the term $\|D_{\text{sym}}[\hat{\sigma}-\widetilde{\sigma}]\|_{L_2}$ we use the following result of Plum (see [24]): if $\sigma\in(H_0^1(\Omega))^2$, then

$$\|D_{\text{sym}}[\sigma]\|_{L_2}^2 \leq \|div\ \sigma\|_{L_2}^2 + \frac{1}{2}\|rot\ \sigma\|_{L_2}^2.$$

This yields now in view of (40)

$$\|D_{\text{sym}}[\widetilde{\sigma} - \hat{\sigma}]\|_{L_2}^2 \leq \|div\,(\hat{\sigma} - \widetilde{\sigma})\|_{L_2}^2 + \frac{1}{2}\|rot\,\widetilde{\sigma}\|_{L_2}^2 \leq \Big(D^2 + \frac{1}{2}\Big)\|rot\,\widetilde{\sigma}\|_{L_2}^2.$$

□

Using the estimates (41), (44) and (45), we obtain a computable upper bound for $\|\omega\|_\infty$.

4.4 Computation of the error bound α

Condition (16) reads in this context

$$\delta \leq \frac{1}{K}\,\alpha - a \cdot (e^{C\alpha} - C\alpha - 1), \tag{77}$$

with

$$a = \lambda\,|\Omega|^{\frac{1}{2}}\,C_{H_0^2 \hookrightarrow L_2}\,e^{\|\omega\|_\infty}.$$

A simple analysis of the right-hand side of (77) as a function of α shows, that if $\lambda > 0$ this function is concave and the maximal value is attained in $\alpha_0 = \frac{1}{C}\log\left(1 + \frac{1}{aKC}\right)$. This means that δ has to fulfil

$$\delta \leq \frac{1}{K}\,\alpha_0 - a\cdot(e^{C\alpha_0} - C\alpha_0 - 1) = a\big((1+r)\log(1+r) - r\big) \quad \text{with } r = \frac{1}{aKC} \tag{78}$$

to obtain an enclosure result. In the affirmative case the minimal α satisfying (77) with the given δ can be obtained by solving (77) with equality instead of inequality approximately e.g. by Newton method, then verifying the inequality a posteriori for a value a little bit larger than the approximate minimum.

4.5 Numerical examples

We investigated the fourth order Gelfand-equation on three different domains: on the unit square, on a disc-like domain $\Omega_\circ$ and on a dumbbell-like domain Ω_d, see Figures 5 and 9. We computed on each domain numerical approximations to $(u, v) \in H_0^1(\Omega) \times H^1(\Omega)$ via Newton method. These numerical solutions behave the same way in some sense on each domain. There is a value $0 < \lambda_0$, such that the Newton method safely converged for all $\lambda < \lambda_0$. For each $\lambda < \lambda_0$ we found (at least) two solutions, a lower and an upper one. Moreover, on the nonconvex dumbbell-like domain Ω_d there are two more nonsymmetric solutions as well, which we will describe below. If we investigate the maximum norm of the lower solutions, we find that it increases as $\lambda \to \lambda_0$. Analogously, the maximum norm of the upper solutions decreases on $]0, \lambda_0]$, and the two branches tends to the same value near to λ_0, see the pictures below. Thus λ_0 seems to be a turning point.

The next question is, how such approximations can be obtained. One starts at the origin, i.e., if $\lambda = 0$ with the trivial solution $u = 0$. Then we step towards the apparent turning point. In every step λ will be chosen a little bit larger, and one uses as starting function for the approximate solver the approximate solution for the previous λ (continuity method). In this way one can get the lower branch. Now, if we multiply a lower solution near to the turning point with an appropriate factor, say 1.5, and we start the approximate solver with this starting value, then we arrive on the upper solution branch. Then the continuity method used at the lower branch can be applied also here, and one can "walk" along the upper solution branch.

There are two crucial values for obtaining enclosure: the defect δ in (25) and the constant K in (27). Calculating δ with the same number of unknowns in the finite element context, we observe, that δ increases on the lower branch if λ increases. It increases further if we bend onto the upper solution branch and we go along this branch with decreasing λ. This means, that it is harder and harder to fulfil (78) as we go along the lower then back on the upper solution branch.

To obtain an upper bound for the constant K according to Section 3.5 we need eigenvalue bounds for the eigenvalues of (67) close to 1. On the lower solution branch all the eigenvalues are smaller than 1. Moreover, the largest of them is getting closer to 1, if λ increases. Thus K increases, which means that the right-hand side of (78) decreases. Therefore it is again harder and harder to fulfil (78), if λ increases. On the upper solution branch one of the eigenvalues slip above 1. As λ decreases from the apparent turning point to 0, the eigenvalue above 1 increases and the second largest eigenvalue (below 1) increases to 1, see Figure 1. This has the consequence, that on the upper solution branch K decreases first, as λ decreases, and then it increases again. One can obtain an enclosure of the solution on the upper branch with the least computing effort near to this minimum of K.

We remark that on the the nonconvex dumbbell-like domain Ω_d for the upper approximate solution branch one finds numerically two, clustered eigenvalues of (67) larger then 1, see the examples below.

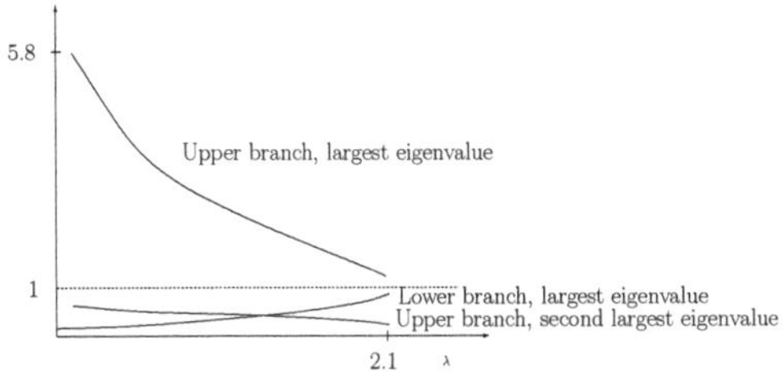

Figure 1: *Eigenvalues of (67) of the Gelfand-equation on* Ω_d

We obtained the following results on the three domains.

1. Unit square

On the unit square the value of the apparent turning point is approximately 437. The maximum norm of both solution branches was about 1.5 near to the turning point, see Figure 2.

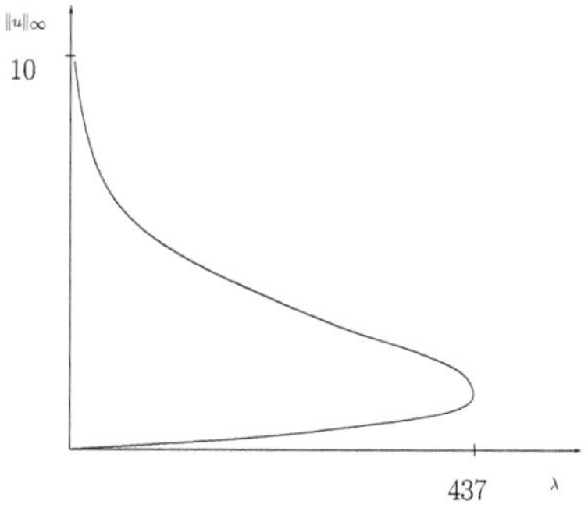

Figure 2: *Maximum norm on the lower and upper solution branches of the Gelfand-equation on the unit square*

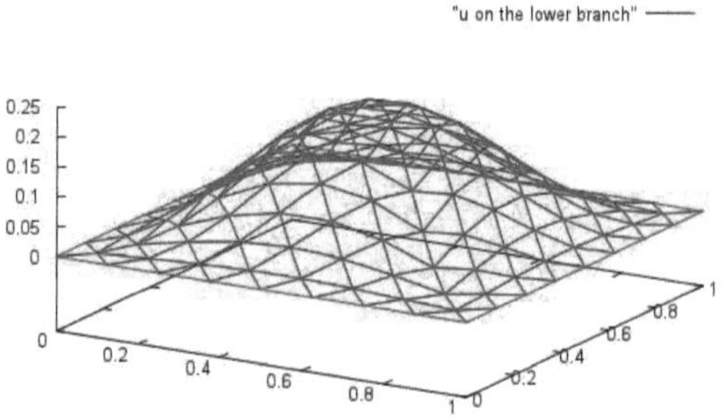

Figure 3: *Numerical solution for* $\lambda = 150$ *on the lower solution branch of the Gelfand-equation on the unit square*

An upper bound for the value of the div-rot constant on the unit square is $D = \sqrt{2} + 1$, see [28].

The first Dirichlet-eigenvalue of the Laplace-operator is known on the unit-square, it is $\mu_1 = 2\pi^2$. Moreover, bounds for the first eigenvalue $\hat{\mu}_1$ of the

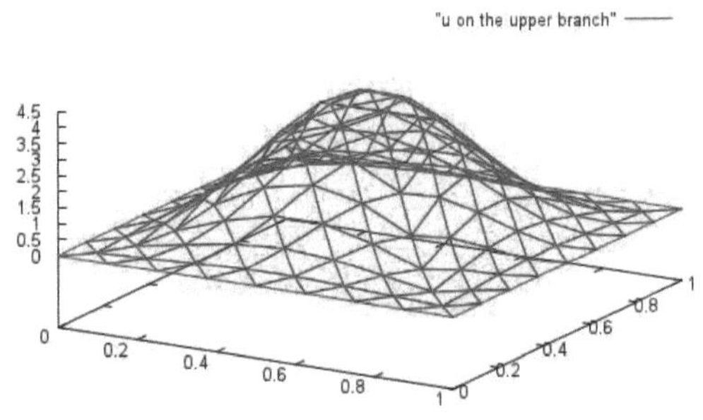

Figure 4: *Numerical solution for $\lambda = 150$ on the upper solution branch of the Gelfand-equation on the unit square*

biharmonic operator are also known, see [31]. It holds

$$\hat{\mu}_1 \in [1294.93394, 1294.933988].$$

Using the results of Section 6.2 this results in the following upper bounds for the required imbedding constants:

$$C_{H_0^1 \hookrightarrow L_2} \leq \frac{1}{\sqrt{2}\pi}, \; C_{H_0^2 \hookrightarrow L_2} \leq 0.0278, \; C_{H_0^1 \hookrightarrow L_4} \leq \frac{1}{\sqrt{2\pi}}, \; C_{H_0^1 \hookrightarrow L_6} \leq \left(\frac{3}{4\sqrt{2}\pi} \right)^{\frac{1}{3}}.$$

For calculating the constant C of (26) we have to determine the constants C_0, C_1, C_2 of Theorem 3.4. The domain Q can be chosen as a square. According to (61) the optimal side length of the square is $\sqrt[4]{\frac{45}{2}\frac{1}{\pi}} < 0.694$, that obviously fulfils condition (50). Then with $\alpha = 3$ we have

$$C_0 \leq 1.7648, \quad C_1 \leq 0.4473, \quad C_2 \leq 0.0896,$$

thus Corollary 3.5 results in

$$C \leq 0.3804.$$

For the upper estimate of $\|\omega\|_\infty$ in (4.7) one needs the constants C_0, C_1, C_2, C_3 of Theorem 3.4. Due to the form of C_3 we choose the side length of the square Q maximal, i.e. $Q = [0,1]^2$. Then with $\alpha = 3$ and $p = 4$ we have

$$C_0 \leq 1.225, \quad C_1 \leq 0.448, \quad C_2 \leq 0.13, \quad C_3 \leq 0.612.$$

Using the above constants we obtained the following enclosure results.

a) Lower solution branch

(i) For $\lambda = 100$ we computed the following defects with 2402 unknowns in the finite element approximations:

$$\begin{aligned}
\|\widetilde{\sigma} - \nabla \widetilde{u}\|_{L_2} &\leq 0.000238661 \\
\|\widetilde{\sigma} - \nabla \widetilde{u}\|_{L_4} &\leq 0.000363857 \\
\|rot\ \widetilde{\sigma}\|_{L_2} &\leq 0.00221956 \\
\|div\ \widetilde{\sigma} + \widetilde{v}\|_{L_2} &\leq 0.00303927 \\
\|\nabla \widetilde{v} - \widetilde{\rho}\|_{L_2} &\leq 0.126857 \\
\|div\ \widetilde{\rho} + F(\widetilde{u})\|_{L_2} &\leq 3.06379 \\
\text{and } \|\omega\|_\infty &\leq 0.1419.
\end{aligned}$$

To obtain an upper bound for the constant K, according to Section 3.5 we need eigenvalue bounds for the eigenvalues of (67) close to 1. In this case all the eigenvalues are smaller than 1, thus we need only an upper bound for the largest eigenvalue, that is approximately

$$\kappa_1 \approx 0.086.$$

According to Sections 6.3.2 and 6.3.5, the inverses of the eigenvalues of problem (145) with $s = 0$ on the unit square yield rough upper bounds for the eigenvalues of (67). Bounds for the eigenvalues of (145) with $s = 0$ on the unit square can be clearly computed from the Dirichlet eigenvalues of the biharmonic operator on the unit square. Very good bounds for the latter eigenvalues are computed in [31]. Thus we obtain

$$\frac{1}{\lambda_1^{(0)}} \leq 0.0896.$$

The eigenvalue $\lambda_1^{(0)}$ is in this case very close to the approximate value of κ_1, since the approximation $\widetilde{u}$ is very "flat", it is everywhere "close" to its maximum value, that is less then 0.14. Moreover, κ_1 is "far away" from 1, since $\lambda = 100$ is still far away from the apparent turning point 437. Therefore the upper bound $\frac{1}{\lambda_1^{(0)}}$ is also less then 1 and it is still enough to obtain enclosure for a solution on the lower solution branch. In this way we obtained $\widetilde{K} \leq 1.09853$. This gives with $\gamma \leq 0.0007248$ that $K \leq 1.0994$. From the defects we computed

$$\delta = 0.191354,$$

that is small enough to fulfil (78). For the minimal α holds $\alpha_{min} \leq 0.2398$. This gives the enclosure of a true solution $u^\star \in H_0^2(\Omega)$

$$\|u^\star - \omega\|_{H_0^2} \leq 0.2398,$$

and in particular

$$\|u^\star - \omega\|_\infty \leq 0.09409.$$

From $\|\widetilde{u} - \omega\|_\infty \leq 0.00264325$ follows then

$$\|u^\star - \widetilde{u}\|_\infty \leq 0.09674.$$

(ii) For $\lambda = 200$ we computed the following defects with 9410 unknowns in the finite element approximations:

$$\begin{aligned}
\|\widetilde{\sigma} - \nabla\widetilde{u}\|_{L_2} &\leq 0.0000674049 \\
\|\widetilde{\sigma} - \nabla\widetilde{u}\|_{L_4} &\leq 0.000106268 \\
\|rot\ \widetilde{\sigma}\|_{L_2} &\leq 0.000539964 \\
\|div\ \widetilde{\sigma} + \widetilde{v}\|_{L_2} &\leq 0.0011045 \\
\|\nabla\widetilde{v} - \widetilde{\rho}\|_{L_2} &\leq 0.0877672 \\
\|div\ \widetilde{\rho} + F(\widetilde{u})\|_{L_2} &\leq 1.21353 \\
\text{and } \|\omega\|_\infty &\leq 0.310377.
\end{aligned}$$

As in the case $\lambda = 100$ we used as rough upper bound for the largest eigenvalue of (67) that is

$$\kappa_1 \approx 0.194,$$

the inverse of the smallest eigenvalue of (145) with $s = 0$, that is

$$\frac{1}{\lambda_1^{(0)}} \leq 0.212.$$

Thus we obtained $\widetilde{K} \leq 1.277778$. This gives with $\gamma \leq 0.000389057$ that $K \leq 1.278414$. From the defects we have

$$\delta = 0.0847791,$$

that is small enough to fulfil (78). For the minimal α holds $\alpha_{min} \leq 0.132799$. This gives the enclosure of the true solution $u^\star \in H_0^2(\Omega)$

$$\|u^\star - \omega\|_{H_0^2} \leq 0.132799,$$

and in particular

$$\|u^\star - \omega\|_\infty \leq 0.0521178.$$

From $\|\widetilde{u} - \omega\|_\infty \leq 0.00066064$ follows then

$$\|u^\star - \widetilde{u}\|_\infty \leq 0.05277844.$$

b) Upper solution branch

For the upper solution branch we computed from the defects the following values for δ and for the approximate maximal value R_{max} of the right-hand side of (78) using the Ritz-approximation of the eigenvalue problem (67) for approximating K.

λ	δ	R_{max}
200	3.9	0.001699
300	2.4	0.004006
350	1.84	0.006027
380	1.54	0.007293
400	1.33	0.004994

(The maximal value for R_{max} depending on λ seems to be near to $\lambda = 380$.)

For these computations we needed 9410 unknowns for the finite element approximations, circa 30 M memory and it took 18 minutes. To obtain a δ smaller than H we would need much more unknowns. If we assume on the basis of our experience, that δ decreases by a factor 1/2 by each refinement of the mesh, i.e. by dividing all triangles into four subtriangles, even in the case $\lambda = 380$ we would need $\lceil \log_2 \frac{1.54}{0.007283} \rceil = 8$ refinements. This means circa $9410 \cdot 4^8 = 616693760$ unknowns, at least $30 \cdot 4^8 M = 1920G$ memory and $18 \cdot 4^8$ min $= 820$ days. This is unfortunately a task for the future computers or different numerical methods.

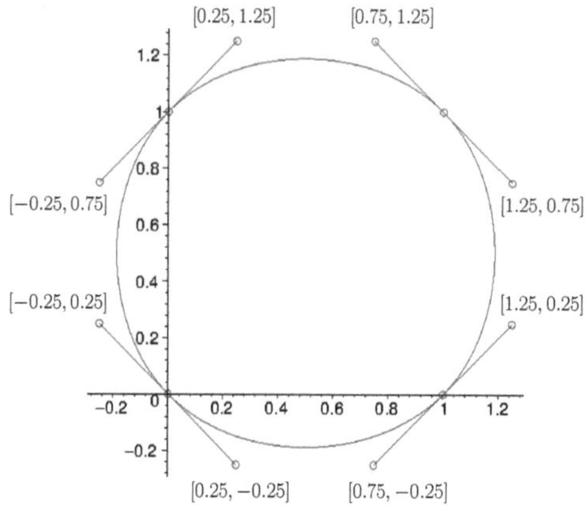

Figure 5: *The disc-like domain* $\Omega_\circ$

2. Disc-like domain $\Omega_\circ$

This domain $\Omega_\circ$ is bordered by four cubic Bézier-splines with corners of the unit square as knots. The control points are determined such that $\Omega_\circ$ is a C^1-domain, see Figure 5. A parametrisation of $\partial\Omega$ is then

$$g_i(t) = (1-t)^3 A_i + 3(1-t)^2 t C_{i1} + 3(1-t)t^2 C_{i2} + t^3 A_{i+1}, \quad i = 1, \ldots, 4, \ t \in [0,1],$$

with

$$A_1 = A_5 = (0,0), \quad A_2 = (1,0), \quad A_3 = (1,1), \quad A_4 = (0,1),$$

$$C_{11} = (0.25, -0.25), \quad C_{12} = (0.75, -0.25), \quad C_{21} = (1.25, 0.25),$$
$$C_{22} = (1.25, 0.75), \quad C_{31} = (0.75, 1.25), \quad C_{32} = (0.25, 1.25),$$
$$C_{41} = (-0.25, 0.75), \quad C_{42} = (-0.25, 0.25).$$

Also in this case we have two approximate solution branches, as described above. The value of the apparent turning point is approximately 151, the maximum norm of the approximate solutions on both branches is about 1.5 near to the turning point, see Figure 6.

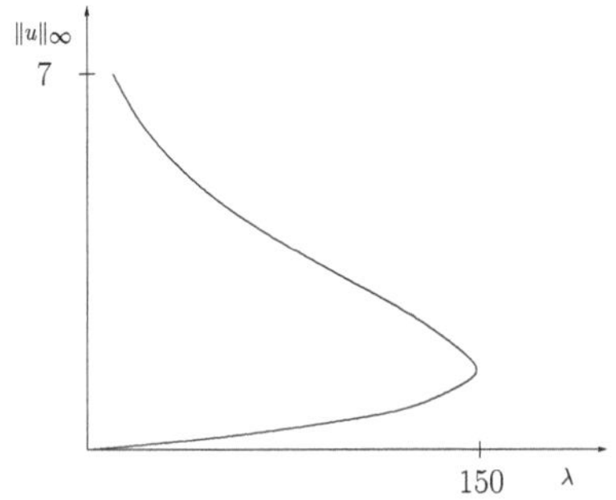

Figure 6: *Maximum norm on the lower and upper solution branches of the Gelfand-equation on* $\Omega_\circ$

A lower bound $\underline{\mu_1}$ for the first Dirichlet-eigenvalue of the Laplace operator on $\Omega_\circ$ is calculated in Section 6.3.4, Example 1: it holds that $\underline{\mu_1} = 11.9122$. Hence according to Section 6.2 we obtain the following upper bounds for the imbedding constants:

$$C_{H_0^1 \hookrightarrow L_2} \leq 0.2898, \quad C_{H_0^2 \hookrightarrow L_2} \leq 0.08395, \quad C_{H_0^1 \hookrightarrow L_4} \leq 0.461, \quad C_{H_0^1 \hookrightarrow L_6} \leq 0.6013.$$

A better bound can be calculated for $C_{H_0^2 \hookrightarrow L_2}$ by using bounds for the first eigenvalue of the biharmonic operator, see Section 6.2. According to the Min-Max Principle a lower bound for the first eigenvalue of the biharmonic operator on a square containing $\Omega_\circ$ yields a lower bound for the first eigenvalue of the biharmonic operator on $\Omega_\circ$. Using the bounds in [31] we obtain $C_{H_0^2 \hookrightarrow L_2} \leq 0.05254$.

An upper bound for the div-rot constant is calculated in Subsection 6.1.1, Example 2, it is $D \leq 1.089$.

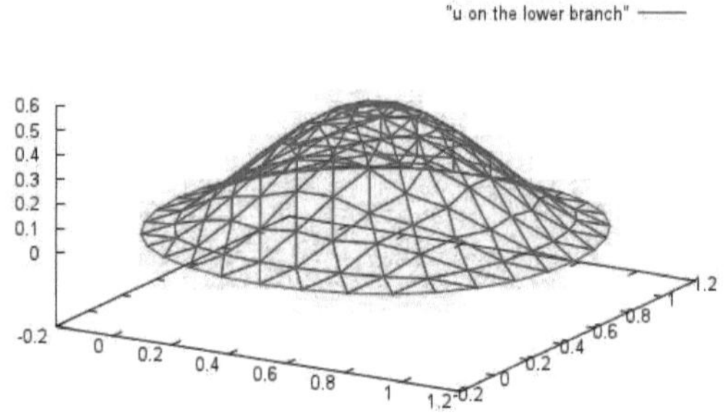

Figure 7: *Numerical solution for* $\lambda = 100$ *on the lower solution branch of the Gelfand-equation on* $\Omega_\circ$

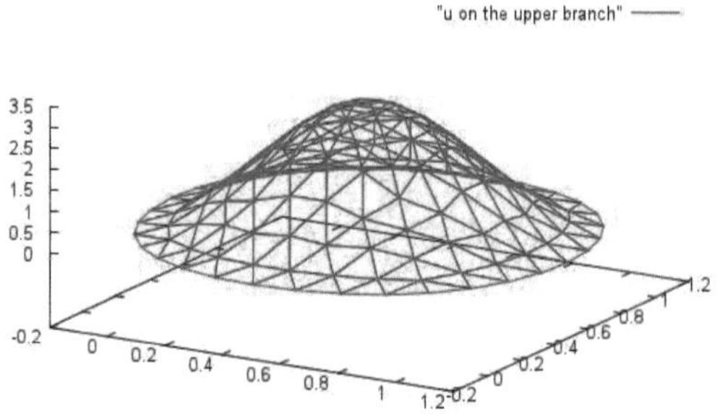

Figure 8: *Numerical solution for* $\lambda = 100$ *on the upper solution branch of the Gelfand-equation on* $\Omega_\circ$

For calculating the constants C_0, C_1, C_2, C_3 of Theorem 3.4, we can choose the domain Q as a circular disc. The optimal radius for minimising the constant C according to (60) is $\sqrt{2\sqrt{\frac{21}{5}\frac{1}{\mu_1}}} > 0.57$, that does not fulfil condition (50). Therefore we choose for ρ the maximal possible radius $\rho = 0.53$ that satisfies

(50). Then one gets from (59) with $\alpha = 3$, $p = 4$ that

$$C_0 \le 1.3041, \quad C_1 \le 0.3786, \quad C_2 \le 0.0894, \quad C_3 \le 0.4583.$$

Using Corollary 3.5 this results in

$$C \le 0.4183.$$

With the above constants we obtained the following enclosure results.

a) Lower solution branch

For $\lambda = 50$ we computed the following defects with 9410 unknowns in the finite element approximations:

$$\begin{aligned}
\|\widetilde{\sigma} - \nabla \widetilde{u}\|_{L_2} &\le 0.000889917 \\
\|\widetilde{\sigma} - \nabla \widetilde{u}\|_{L_4} &\le 0.00679581 \\
\|rot\ \widetilde{\sigma}\|_{L_2} &\le 0.00347851 \\
\|div\ \widetilde{\sigma} + \widetilde{v}\|_{L_2} &\le 0.0127502 \\
\|\nabla \widetilde{v} - \widetilde{\rho}\|_{L_2} &\le 0.310339 \\
\|div\ \widetilde{\rho} + F(\widetilde{u})\|_{L_2} &\le 1.38171 \\
\text{and } \|\omega\|_\infty &\le 0.216446.
\end{aligned}$$

As in the case of the unit square we used a rough upper bound for the largest eigenvalue of (67)

$$\kappa_1 \approx 0.134.$$

In this case we cannot use the inverse of the smallest eigenvalue of (145) with $s = 0$ on $\Omega_\circ$, since bounds for this eigenvalue are not known. Therefore let us consider (145) with $s = 0$ on the square

$$\Omega_0 = \operatorname{conv}\left(\left(-\frac{3}{16}, -\frac{3}{16}\right), \left(\frac{19}{16}, -\frac{3}{16}\right), \left(\frac{19}{16}, \frac{19}{16}\right), \left(-\frac{3}{16}, \frac{19}{16}\right)\right)$$

that contains $\Omega_\circ$. This eigenvalue problem can be chosen as a comparison problem for (145) with $s = 0$ on $\Omega_\circ$, since condition (111) is fulfilled with

$$H_0 = H_0^2(\Omega_0), \qquad \langle u, v\rangle_{H_0} = \int_{\Omega_0} \Delta u \Delta v + \alpha uv \, dx,$$

$$H = \{u \in H_0^2(\Omega_0) \colon u = 0 \text{ on } \Omega_0 \setminus \Omega_\circ\}, \qquad \langle u, v\rangle_H = \int_{\Omega_\circ} \Delta u \Delta v + \alpha uv \, dx,$$

and

$$N_0(u, \varphi) = \int_{\Omega_0} (M + \alpha) u\varphi \, dx \quad \text{and} \quad N(u, \varphi) = \int_{\Omega_\circ} (\widetilde{c}(x) + \alpha) u\varphi \, dx,$$

where we used the notations of Sections 6.3.2 and 6.3.5. Again with the help of the bounds in [31] we obtained the following upper bound for the inverse of the smallest eigenvalue of (145) with $s = 0$ on Ω_0

$$\frac{1}{\lambda_1^{(0)}} \leq 0.478.$$

We can observe that this bound has larger distance to the approximate value of κ_1 as in the case of the unit square, because we did two steps now in the estimate: we enlarged the left-hand side N and we also enlarged the domain. But fortunately, this rough upper bound is still "much" less then 1 and that is why it is still enough to obtain an enclosure of a solution.

This way we obtained $\widetilde{K} \leq 1.153733$, and with $\gamma \leq 0.00154927$ we have $K \leq 1.155799$. From the defects we computed

$$\delta = 0.22622,$$

that is small enough to fulfil (78). For the minimal α holds $\alpha_{min} \leq 0.341$. This gives the enclosure of the true solution $u^\star \in H_0^2(\Omega)$

$$\|u^\star - \omega\|_{H_0^2} \leq 0.3410,$$

and in particular

$$\|u^\star - \omega\|_\infty \leq 0.1427.$$

From $\|\widetilde{u} - \omega\|_\infty \leq 0.00589445$ follows then

$$\|u^\star - \widetilde{u}\|_\infty \leq 0.1486.$$

b) Upper solution branch

On the upper solution branch we computed the following values for δ and for the approximate maximal value R_{max} of the right-hand side of (78) using Ritz-approximations of the eigenvalue problem (67) for approximating K. For these computations we used 9410 unknowns in finite element context.

λ	δ	R_{max}
50	4.0207	0.000812
100	1.9204	0.004248
120	1.5103	0.007099

Due to similar arguments as in the case of the unit square we could not achieve enclosure on this branch.

3. Dumbbell-like domain Ω_d

Also this domain Ω_d is bordered by cubic Bézier-splines such that Ω_d is a C^1-domain, see Figure 9. Now the base points are

$$A_1 = A_{11} = (0,0),\ A_2 = (2,2),\ A_3 = (4,1),\ A_4 = (6,1),\ A_5 = (8,2),$$

$$A_6 = (10,0),\ A_7 = (8,-2),\ A_8 = (6,-1),\ A_9 = (4,-1),\ A_{10} = (2,-2),$$

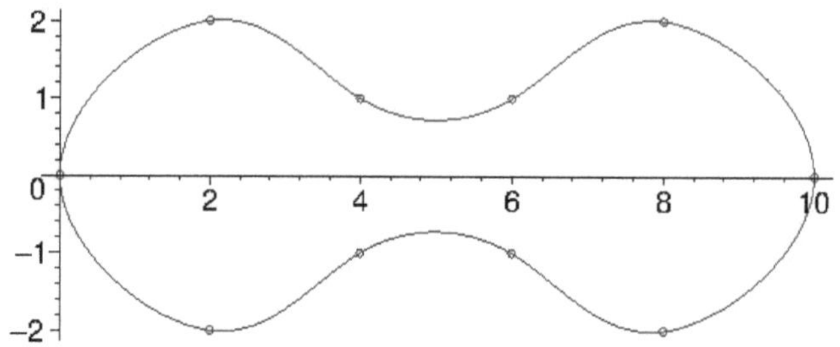

Figure 9: *The dumbbell-like domain* Ω_d

and the so-called de Boor points are

$$B_1 = B_{11} = \left(\frac{-22}{19}, 0\right),\ B_2 = \left(\frac{44}{19}, \frac{54}{19}\right),\ B_3 = \left(\frac{74}{19}, \frac{12}{19}\right),\ B_4 = \left(\frac{116}{19}, \frac{12}{19}\right),$$

$$B_5 = \left(\frac{146}{19}, \frac{54}{19}\right),\ B_6 = \left(\frac{212}{19}, 0\right),\ B_7 = \left(\frac{146}{19}, -\frac{54}{19}\right),\ B_8 = \left(\frac{116}{19}, -\frac{12}{19}\right),$$

$$B_9 = \left(\frac{74}{19}, -\frac{12}{19}\right),\ B_{10} = \left(\frac{44}{19}, -\frac{54}{19}\right).$$

The control points are determined by $C_{i1} = \frac{2}{3}B_i + \frac{1}{3}B_{i+1}$, $C_{i2} = \frac{1}{3}B_i + \frac{2}{3}B_{i+1}$. A parametrisation of $\partial\Omega$ is then

$$g_i(t) = (1-t)^3 A_i + 3(1-t)^2 t C_{i1} + 3(1-t)t^2 C_{i2} + t^3 A_{i+1},\quad i = 1, \dots, 10,\ t \in [0, 1].$$

As already mentioned, there exist four approximate solution branches, two symmetric branches (an upper and a lower) and two nonsymmetric ones. The two nonsymmetric solution branches are reflections of each other to the symmetry-axis $x = 5$ of Ω_d, see Figures 10, 11, 12 and 13.

The solutions on the lower symmetric branch are getting more flat near to the origin and more high near to the apparent turning point, that is approxi-

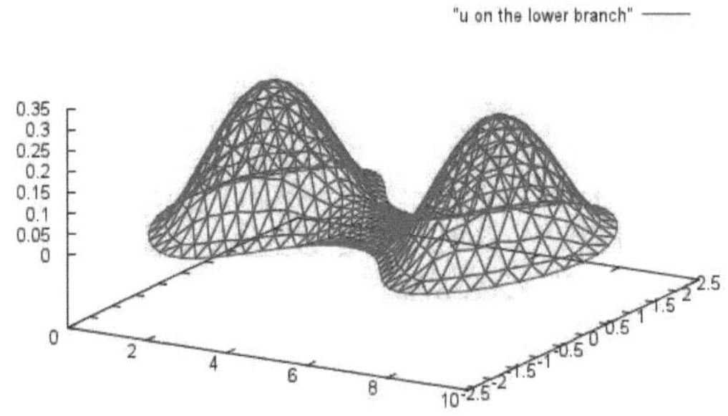

Figure 10: *Numerical solution for* $\lambda = 1$ *on the lower solution branch of the Gelfand-equation on* Ω_d

mately 2.12. The upper symmetric solutions are vice versa, larger near to the origin and getting lower (as high as the lower solutions) near to the apparent turning point. The upper branch seems to be unbounded near to 0. One can see this behaviour, this nose-shaped curve in Figure 14.

The nonsymmetric approximate solutions are getting more "symmetric" near to a apparent bifurcation point close to the apparent turning point. This phenomenon can be observed in Figure 14. Curve L shows the maximum of the nonsymmetric solutions on the left-half of the domain (i.e., for $x \leq 5$), curve R shows the maximum on the right-half of the domain (i.e., for $x \geq 5$). (Of course, for the other nonsymmetric branch we would get the same picture with L instead of R, and vice versa.) The nonsymmetric branches also seem to be unbounded near to 0. Observe the similarity of the curves on the subfigures of Figure 14.

One can relatively easily find the two symmetric branches as described at the beginning of this section. Getting the nonsymmetric branches is a bit more costly. One can make use of the following technique: one perturbs the domain a little bit, destroying its symmetry. Then the above mentioned continuity

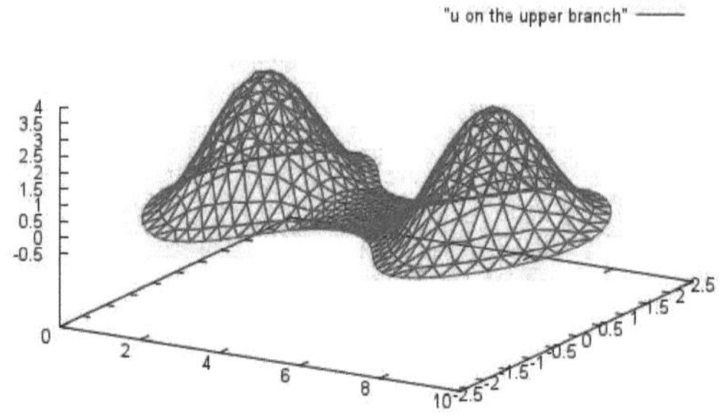

Figure 11: *Numerical solution for* $\lambda = 1$ *on the upper solution branch of the Gelfand-equation on* Ω_d

method can be applied: start at the point $\lambda = 0$ and then "walk" upwards until the approximate solver method converges. In this way one obtains nonsymmetric solutions on the nonsymmetric domain. Now one chooses a λ and builds a rough approximation on the symmetric domain, which is similar to the nonsymmetric approximation to the same λ on the nonsymmetric domain. The approximate solver will be started with this rough nonsymmetric approximation. Then it converges most likely to a nonsymmetric solution on the symmetric domain. The last task is to reflect this nonsymmetric solution with respect to the symmetry-axis $x = 5$ to get the other nonsymmetric approximate solution. (Or we can start the approximate solver with the reflected rough nonsymmetric approximation.)

If we consider approximate eigenvalues of the eigenvalue problem (67) on the upper solution branch (assuming that these approximations are good enough), then we can assert that the two biggest eigenvalues over 1 are clustered. For example for $\lambda = 1$ we have

$$\kappa_1 \approx 2.74038, \quad \kappa_2 \approx 2.74036, \quad \kappa_3 \approx 0.4578.$$

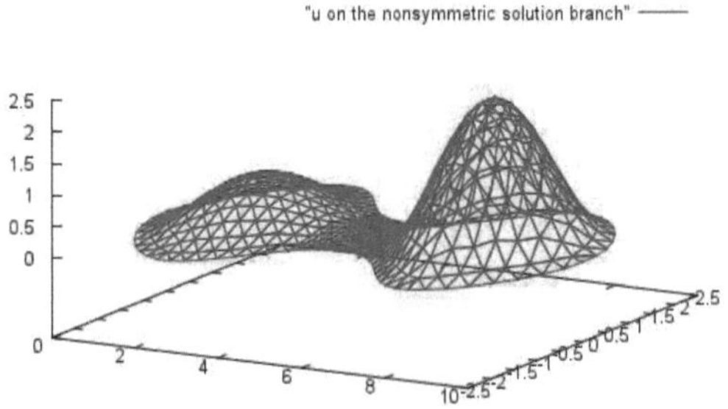

Figure 12: *Numerical solution for* $\lambda = 1.8$ *on the nonsymmetric solution branch of the Gelfand-equation on* Ω_d

This phenomenon may point to the fact, that the turning point is not simple, or that there is a bifurcation point (where unsymmetric solutions bifurcate from the symmetric branch) very close to the turning point. The investigation of this question requires other methods.

For the method of Plum we have to determine a lower bound for the smallest Dirichlet eigenvalue of the Laplace operator, the div-rot constant, and the constants C_0, C_1, C_2 and C_3 from Theorem 3.4.

A lower bound for the smallest eigenvalue of the Laplace operator is established in Section 6.3.4, it is $\underline{\mu_1} = 1.3819$.

An upper bound for the div-rot constant is calculated in Section 6.1.1, Example 3, it is $D \leq 87.56$.

For determining the constants C_0, C_1, C_2 and C_3 from Theorem 3.4 we choose Q as a circular disc. We obtained by curvature calculations that the radius of the maximal disc that fulfils (50), satisfies $r \leq 0.72$. The optimal radius for calculating the constant C according to (60) is larger than 1.7. Thus we

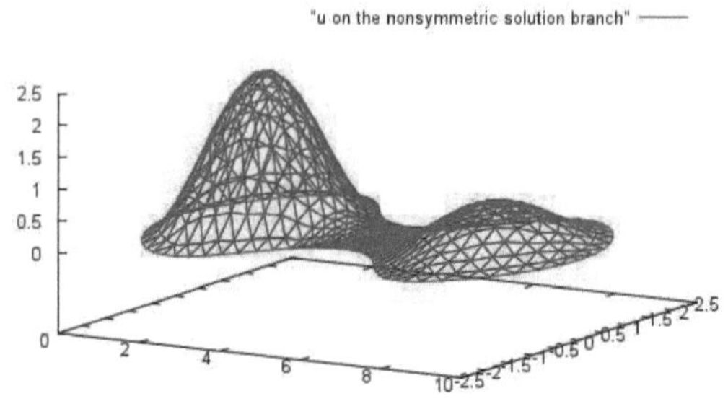

Figure 13: *Numerical solution for* $\lambda = 1.8$ *on the nonsymmetric solution branch of the Gelfand-equation on* Ω_d

choose $r = 0.72$. Then according to (59) this gives the following values with $\alpha = 3$, $p = 4$:

$$C_0 \leq 0.9600, \quad C_1 \leq 0.3786, \quad C_2 \leq 0.1215, \quad C_3 \leq 0.5342.$$

Using Corollary 3.5 this results in

$$C \leq 1.4602.$$

a) Lower solution branch

For $\lambda = 0.4$ we computed the following defects with 32738 unknowns in the

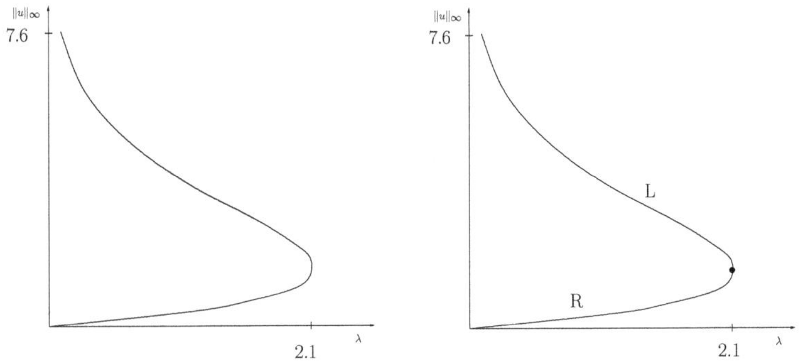

Figure 14: *Maximum norm of the numerical solutions of the symmetric branches and on the nonsymmetric branches of the Gelfand-equation on* Ω_d

finite element approximations:

$$\begin{aligned}
\|\widetilde{\sigma} - \nabla \widetilde{u}\|_{L_2} &\leq 0.000423 \\
\|\widetilde{\sigma} - \nabla \widetilde{u}\|_{L_4} &\leq 0.00202 \\
\|rot\ \widetilde{\sigma}\|_{L_2} &\leq 0.0006243 \\
\|div\ \widetilde{\sigma} + \widetilde{v}\|_{L_2} &\leq 0.003008 \\
\|\nabla \widetilde{v} - \widetilde{\rho}\|_{L_2} &\leq 0.0211647 \\
\|div\ \widetilde{\rho} + F(\widetilde{u})\|_{L_2} &\leq 0.00969 \\
\text{and } \|\omega\|_\infty &\leq 0.192763.
\end{aligned}$$

Since the eigenvalues of the biharmonic operator are not known on Ω_d, we can not simply use as rough upper bound the inverse of the smallest eigenvalue of (145) with $s = 0$ to obtain an upper bound for $\widetilde{K}$, as in case of the unit square. One may think of using the eigenvalues of the biharmonic operator on a square that contains Ω_d, but they are too small to get an enclosure of the solutions of (73) on Ω_d. We should enclose the eigenvalues of (126), or equivalently the eigenvalues of (145) with $s = 1$ on Ω_d with the help of a homotopy as described at the end of Section 6.3.5, (or at least enclose the eigenvalues of an appropriate intermediate problem). But unfortunately the defects in the estimates (134) - (137) converge too slowly, and thus produce matrices A_0, A_1, A_2 with too wide interval entries to get such an enclosure.

As an example let us consider the first step of the domain homotopy. In the course of this homotopy we are aiming at bounds for the smallest eigenvalue of the biharmonic operator on Ω_d, since that would be sufficient to obtain an enclosure of the solutions of (73) on Ω_d. The two smallest eigenvalues are clustered, it holds that

$$\hat{\lambda}_1 \approx 6.30 \quad \text{and} \quad \hat{\lambda}_2 \approx 6.31.$$

Since bounds for the eigenvalues of the biharmonic operator are known on squares, we can start the domain homotopy with the square

$$\Omega_0 = \text{conv}((0,0),(10,0),(10,10),(0,10))$$

that contains Ω_d. We can choose for β in the first homotopy step the fourteenth eigenvalue $\lambda_{14} \geq 8.78$ of the biharmonic operator on Ω_0 and we are aiming at the enclosure of the thirteenth eigenvalue $\lambda_{13} \approx 8.74$ of the biharmonic operator on the rectangle $\text{conv}((0,0),(10,0),(10,8.87),(0,8.87))$. We demonstrate with the example of the interval matrices on the left- and on the right-hand side of (143), why the enclosure failed in this case, see the following table:

# unknowns	$\overline{A_0} - \beta\overline{A_1}$	$\overline{A_0} - 2\beta\overline{A_1} + \beta^2\overline{A_2}$
2402	[-5.097296, 5.033425]	[-22.542253, 25.472481]
9410	[-0.650203, 0.586540]	[-6.949375, 9.874456]
37250	[-0.110896, 0.047250]	[-5.487982, 8.413077]
148226	[-0.044654, -0.018991]	[-5.315913, 8.241028]

Since the matrix $\overline{A_0} - 2\beta\overline{A_1} + \beta^2\overline{A_2}$ on the right-hand side of (143) contains 0, the interval matrix eigenvalue problem (143) is meaningless. This means that we cannot even get an enclosure in the first homotopy step.

But if we take into account that the Rayleigh-Ritz approximations yield very good approximations to the eigenvalues of (126) (or equivalently the eigenvalues of (145) with $s = 1$), then we can use them to calculate K and α approximately. This will of course not provide a rigorous proof of the existence of true solutions near to the numerical one. But it indicates that if we will be able to enclose the eigenvalues we are looking for, possibly with another method, then the method of Plum will yield us the desired enclosure of a true solution of (73).

From the approximation of the smallest eigenvalue of (126)

$$\nu_1 \approx 5.05$$

we obtained 1.2482 as an approximate upper bound for $\widetilde{K}$. With $\gamma \leq 0.0224$ we get 1.284 as an approximate upper bound for K.

From the defects we have

$$\delta = 0.06522,$$

that is small enough to fulfil (78) (with approximate upper bound for K). For the approximate minimal α holds $\alpha_{min} \approx 0.1384$. If we knew the existence of a true solution $u^\star \in H_0^2(\Omega)$ near to the numerical one then taking

$$\|\widetilde{u} - \omega\|_\infty \leq 0.08145$$

into account we would get the approximate upper bound 0.2689 for $\|u^\star - \widetilde{u}\|_\infty$.

b) Upper solution branch

Again for the upper solution branch we computed the following values for δ and for the approximate maximal value R_{max} of the right-hand side of (78) using the Ritz-approximation of the eigenvalue problem (67) for approximating K. We used for these calculations 32738 unknowns in the finite element approximations.

λ	δ	R_{max}
0.5	44.27	0.00008
1.0	19.2	0.00022
1.5	9.83	0.00049
2.0	4.08	0.00042

Due to similar arguments as in the case of the unit square we could not achieve enclosure on this branch.

c) Nonsymmetric solution branches

Due to similar arguments as in the case of the upper solution branch we could not achieve enclosure on these branches.

5 Application to the Emden-equation

As a second example let us consider the fourth order Emden-equation

$$\Delta^2 u = u^2 \qquad \text{on } \Omega,$$

$$u = \frac{\partial u}{\partial \nu} = 0 \quad \text{on } \partial\Omega.$$

We embed this equation into the following family of problems

$$\Delta^2 u = u^2 + \lambda \qquad \text{on } \Omega, \tag{79}$$

$$u = \frac{\partial u}{\partial \nu} = 0 \quad \text{on } \partial\Omega,$$

where $\lambda \geq 0$. As in case of the Gelfand-equation we first give a computable upper bound for the terms δ and γ, and then we show a possible appropriate choice for the function G. Afterwards we illustrate our method with numerical enclosure results for the Emden-equation on the unit square, on the disc-like domain $\Omega_\circ$ and on the dumbbell-like domain Ω_d.

5.1 Computation of δ

Using the results of Section 3.3 we are left to find a computable constant C_F such that

$$\|\mathcal{F}(\widetilde{u}) - \mathcal{F}(\omega)\|_{H^{-2}} \leq C_F \|\nabla\omega - \nabla\widetilde{u}\|_{L_2}. \tag{80}$$

5.1 Lemma *The constant C_F defined by*

$$C_F = C_{H_0^2 \hookrightarrow L_2} C^2_{H_0^1 \hookrightarrow L_4} \left(\|\widetilde{\sigma} + \nabla\widetilde{u}\|_{L_2} + C_{H_0^1 \hookrightarrow L_2} \cdot \sqrt{D^2+1} \cdot \|rot\ \tilde{\sigma}\|_{L_2} \right)$$

fulfils (80).

Proof: By Hölder's inequality one gets

$$\begin{aligned}
\|\mathcal{F}(\widetilde{u}) - \mathcal{F}(\omega)\|_{H^{-2}} &= \sup_{\varphi \in H_0^2(\Omega), \varphi \neq 0} \frac{|\int_\Omega (\widetilde{u}^2 - \omega^2)\varphi \, dx|}{\|\varphi\|_{H_0^2}} \\
&\leq \sup_{\varphi \in H_0^2(\Omega), \varphi \neq 0} \frac{\|\widetilde{u} - \omega\|_{L_4} \|\widetilde{u} + \omega\|_{L_4} \|\varphi\|_{L_2}}{\|\varphi\|_{H_0^2}} \\
&\leq C_{H_0^2 \hookrightarrow L_2} \cdot C_{H_0^1 \hookrightarrow L_4} \|\widetilde{u} + \omega\|_{L_4} \|\nabla\omega - \nabla\widetilde{u}\|_{L_2}.
\end{aligned}$$

Moreover,

$$\|\widetilde{u} + \omega\|_{L_4} \leq C_{H_0^1 \hookrightarrow L_4} (\|\nabla\widetilde{u} + \widetilde{\sigma}\|_{L_2} + \|\nabla\omega - \widetilde{\sigma}\|_{L_2}).$$

Then by (43) we obtain

$$C_F = C_{H_0^2 \hookrightarrow L_2} C^2_{H_0^1 \hookrightarrow L_4} \left(\|\widetilde{\sigma} + \nabla\widetilde{u}\|_{L_2} + C_{H_0^1 \hookrightarrow L_2} \cdot \sqrt{D^2+1} \cdot \|rot\ \tilde{\sigma}\|_{L_2} \right).$$

□

5.2 Computation of γ

We show how we can obtain in case of the Emden-equation the constant γ satisfying

$$\|(\widetilde{L}-L)[u]\|_{H^{-2},\alpha} \leq \gamma \|u\|_{H_0^2,\alpha} \qquad \text{for all } u \in H_0^2(\Omega). \tag{81}$$

5.2 Lemma *The constant γ defined by*

$$\gamma = 2 \cdot C^2_{H_0^2 \hookrightarrow L_4} C_{H_0^1 \hookrightarrow L_2} \cdot \left(C_{H_0^1 \hookrightarrow L_2} \cdot \sqrt{D^2+1} \cdot \|rot\, \widetilde{\sigma}\|_{L_2} + \|\widetilde{\sigma} - \nabla \widetilde{u}\|_{L_2} \right)$$

fulfils (81).

Proof: By Hölder's inequality one gets

$$\begin{aligned}
\|(\widetilde{L}-L)[u]\|_{H^{-2},\alpha} &= \|F'(\widetilde{u})[u] - F'(\omega)[u]\|_{H^{-2},\alpha} \\
&= 2 \cdot \|(\widetilde{u}-\omega)u\|_{H^{-2},\alpha} \\
&= 2 \cdot \sup_{\varphi \in H_0^2(\Omega),\, \varphi \neq 0} \frac{|\int_\Omega (\widetilde{u}-\omega)u\varphi \, dx|}{\|\varphi\|_{H_0^2,\alpha}} \\
&\leq 2 \cdot \sup_{\varphi \in H_0^2(\Omega),\, \varphi \neq 0} \frac{\|\widetilde{u}-\omega\|_{L_2} \|u\|_{L_4} \|\varphi\|_{L_4}}{\|\varphi\|_{H_0^2}} \\
&\leq 2 \cdot C^2_{H_0^2 \hookrightarrow L_4} \|\widetilde{u}-\omega\|_{L_2} \|u\|_{H_0^2} \\
&\leq 2 \cdot C^2_{H_0^2 \hookrightarrow L_4} \|\widetilde{u}-\omega\|_{L_2} \|u\|_{H_0^2,\alpha}.
\end{aligned}$$

Using estimate (45) for the term $\|\widetilde{u}-\omega\|_{L_2}$ we get the desired upper bound γ. □

5.3 Determination of the function G

In case of the Emden-equation we can easily determine the function G as follows.

5.3 Lemma *The function*

$$G(y) = |\Omega|^{\frac{1}{2}} C_{H_0^2 \hookrightarrow L_2} y^2$$

satisfies (28) and (29).

Proof: According to Section 3.6 let us calculate

$$|F(\omega(x)+y) - F(\omega(x)) - F'(\omega(x))y| = |(\omega(x)+y)^2 - \omega(x)^2 - 2\omega(x)y| = y^2.$$

Thus let $\widetilde{G}(y) = y^2$. One can easily see, that $\widetilde{G}$ is non-decreasing and it satisfies $\widetilde{G}(t) = o(t)$ for $t \to 0+$. Thus according to Section 3.6 the above function G fulfils (28) and (29). □

5.4 Computation of the error bound α

Condition (16) reads in this context

$$\delta \leq \frac{1}{K}\,\alpha - a \cdot (C\alpha)^2, \tag{82}$$

with

$$a = |\Omega|^{\frac{1}{2}}\, C_{H_0^2 \hookrightarrow L_2}.$$

The right-hand side of (82) as a function of α is concave and the maximal value is attained in $\alpha_0 = \frac{1}{2aKC^2}$. This means that δ has to fulfil inequality

$$\delta \leq \frac{1}{K}\,\alpha_0 - a \cdot (C\alpha_0)^2 = \frac{1}{4aK^2C^2}$$

to obtain an enclosure result. In the affirmative case the minimal α is

$$\alpha_{min} = \frac{1 - \sqrt{1 - 4aK^2C^2\delta}}{2aKC^2} = \frac{2K\delta}{1 + \sqrt{1 - 4aK^2C^2\delta}}.$$

5.5 Numerical examples

We investigated the fourth order Emden-equation on the same three domains, as the Gelfand-equation: the unit square, the disc-like domain $\Omega_\circ$ and the dumbbell-like domain Ω_d. Most of the general observations from the beginning of Section 4 about the numerical solutions and about the eigenvalues of (67) are valid here as well.

An essential difference compared to the Gelfand-equation is that the upper branch, as well as the nonsymmetric branches in case of the dumbbell-like domain Ω_d, *seem to be bounded* near to 0 and they reach the case $\lambda = 0$. This means that we found at least one nontrivial approximate solution for the true Emden-equation, i.e., for the case $\lambda = 0$, and moreover on the nonconvex dumbbell-like domain Ω_d we found two more nonsymmetric approximate solutions (which are reflections of each other to the symmetry-axis $x = 5$).

For the values of the constants C_0, C_1, C_2 and C_3, as well as for the imbedding constants and the div-rot constant on the above domains we refer to Section 4. The enclosure results we obtained for the Emden-equation are the following:

1. Unit square

On the unit square the value of the apparent turning point is approximately 349000. The maximum norm of both solution branches was about 900 near to the apparent turning point, see Figure 15.

Using the constants calculated in Section 4 we obtained the following enclosure results.

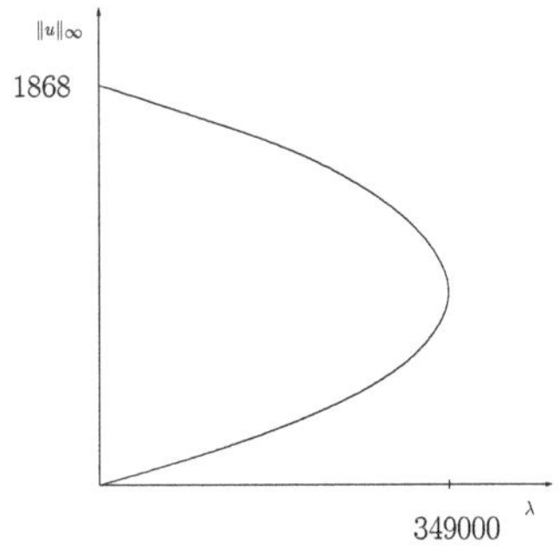

Figure 15: *Maximum norm on the lower and upper solution branches of the Emden-equation on the unit square*

a) Lower solution branch

(i) For $\lambda = 200000$ we computed the following defects with 148226 unknowns in the finite element approximations:

$$\begin{aligned}
\|\widetilde{\sigma} - \nabla\widetilde{u}\|_{L_2} &\leq 0.00105338\\
\|\widetilde{\sigma} - \nabla\widetilde{u}\|_{L_4} &\leq 0.00165689\\
\|rot\ \widetilde{\sigma}\|_{L_2} &\leq 0.018161\\
\|div\ \widetilde{\sigma} + \widetilde{v}\|_{L_2} &\leq 0.031596\\
\|\nabla\widetilde{v} - \widetilde{\rho}\|_{L_2} &\leq 12.7189\\
\|div\ \widetilde{\rho} + F(\widetilde{u})\|_{L_2} &\leq 0.0680583\\
\text{and } \|\omega\|_\infty &\leq 309.506.
\end{aligned}$$

To obtain an upper bound for the constant K according to Section 3.5 we need eigenvalue bounds for the eigenvalues of (67) close to 1. In this case all the eigenvalues are smaller than 1, thus we need only an upper bound for the largest eigenvalue that is approximately

$$\kappa_1 \approx 0.348.$$

As in the case of the Gelfand-equation we used

$$\frac{1}{\lambda_1^{(0)}} \leq 0.481,$$

the inverse of the smallest eigenvalue of problem (145) with $s = 0$ on the unit square as rough upper bounds for κ_1. Now again we are in the lucky situation, that the largest eigenvalue for the comparison problem is still less then 1 and it is still enough to obtain enclosure for the solution on the lower solution branch.

In this way we get $\widetilde{K} \leq 1.917432$. This gives with $\gamma \leq 0.00004259$ that $K \leq 1.917588$. From the defects we computed

$$\delta = 3.01643,$$

that is small enough to fulfil (82). For the minimal α holds $\alpha_{min} \leq 6.3965$. This gives the enclosure of the true solution $u^\star \in H_0^2(\Omega)$

$$\|u^\star - \omega\|_{H_0^2} \leq 6.3965,$$

and in particular

$$\|u^\star - \omega\|_\infty \leq 2.511.$$

From $\|\widetilde{u} - \omega\|_\infty \leq 0.3379$ follows then

$$\|u^\star - \widetilde{u}\|_\infty \leq 2.849.$$

b) Upper solution branch

(i) For $\lambda = 200000$ we computed the following defects with 591362 unknowns in the finite element approximations:

$$\begin{aligned}
\|\widetilde{\sigma} - \nabla\widetilde{u}\|_{L_2} &\leq 0.000629005 \\
\|\widetilde{\sigma} - \nabla\widetilde{u}\|_{L_4} &\leq 0.000933744 \\
\|rot\ \widetilde{\sigma}\|_{L_2} &\leq 0.00898193 \\
\|div\ \widetilde{\sigma} + \widetilde{v}\|_{L_2} &\leq 0.0156803 \\
\|\nabla\widetilde{v} - \widetilde{\rho}\|_{L_2} &\leq 11.0626 \\
\|div\ \widetilde{\rho} + F(\widetilde{u})\|_{L_2} &\leq 0.144046 \\
\text{and } \|\omega\|_\infty &\leq 1527.87.
\end{aligned}$$

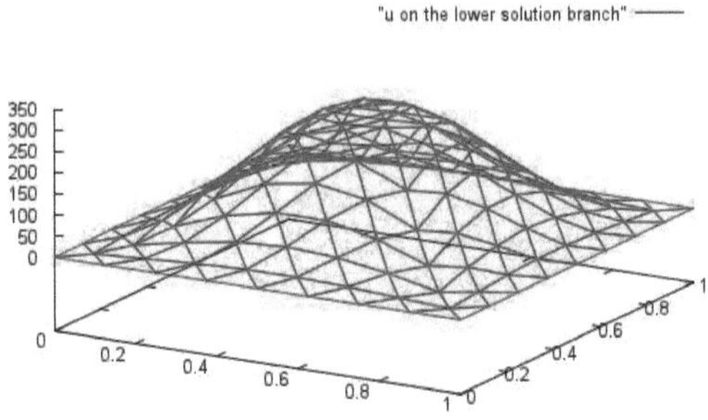

Figure 16: *Numerical solution for* $\lambda = 200000$ *on the lower solution branch of the Emden-equation on the unit square*

To obtain an upper bound for the constant K we need bounds for the eigenvalues κ_i near to 1 of the eigenvalue problem (67), or equivalently bounds for the eigenvalues $\lambda_i = \frac{1}{\kappa_i}$ near to 1 of the eigenvalue problem (145) with $s = 1$ $(i = 1, 2)$. The Rayleigh-Ritz approximations of the two smallest eigenvalues

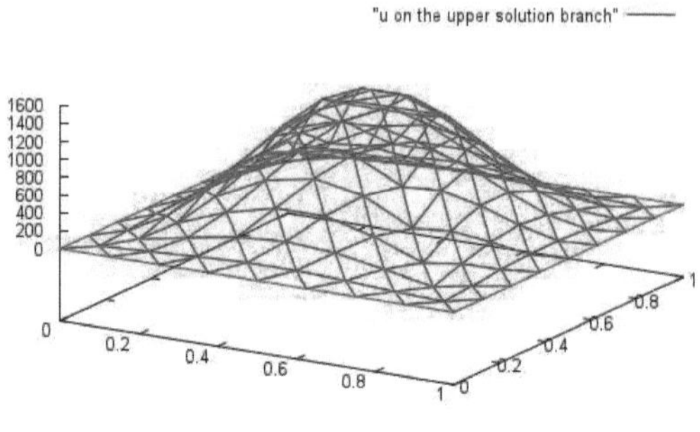

Figure 17: *Numerical solution for* $\lambda = 200000$ *on the upper solution branch of the Emden-equation on the unit square*

are

$$\lambda_1 \approx 0.606 \quad \text{and} \quad \lambda_2 \approx 3.316.$$

The smallest eigenvalue is now smaller than 1, since we are on the upper solution branch. Therefore we need an upper bound for λ_1 and a lower bound for λ_2. For λ_1 we used verified Rayleigh-Ritz upper bound

$$\lambda_1 \leq \overline{\lambda}_1 = 0.61$$

(see Section 6.3.5). To bound λ_2 we calculated the following lower bounds for the two smallest eigenvalues of the eigenvalue problem (145) with $s = 0$ on the unit square

$$\lambda_1^{(0)} \geq 0.42, \quad \text{and} \quad \lambda_2^{(0)} \geq 1.749.$$

The value $\lambda_2^{(0)}$ is bigger than 1, but it is "far away" from the approximate value for λ_2. Nevertheless, it is again sufficient to consider this comparison problem to obtain an enclosure for a solution of (79) and we do not need a homotopy (as described in Section 6.3.5). Thus for λ_2 we used 1.749 as a rough lower bound.

In this way we get $\widetilde{K} \leq 2.335114$. This gives with $\gamma \leq 0.0000214558$ that $K \leq 2.335231$. From the defects we computed

$$\delta = 2.79954,$$

that is small enough to fulfil (82). For the minimal α holds $\alpha_{min} \leq 7.6$. This gives the enclosure of the true solution $u^\star \in H_0^2(\Omega)$

$$\|u^\star - \omega\|_{H_0^2} \leq 7.6,$$

and in particular

$$\|u^\star - \omega\|_\infty \leq 2.99.$$

From $\|\widetilde{u} - \omega\|_\infty \leq 0.0102$ follows then

$$\|u^\star - \widetilde{u}\|_\infty \leq 3.0002.$$

We are left to show that this solution on the upper branch is different from the solution on the lower branch. Let us denote by u_1^* the true solution of (79) with $\lambda = 200000$ on the lower solution branch and by u_2^* on the upper solution branch, and analogously by $\widetilde{u}_1$ and $\widetilde{u}_2$ the corresponding approximations. Then we have from the above results that

$$\|u_2^* - u_1^*\|_\infty \geq \|\widetilde{u}_2 - \widetilde{u}_1\|_\infty - \|u_1^* - \widetilde{u}_1\|_\infty - \|u_2^* - \widetilde{u}_2\|_\infty$$

$$\geq \|\widetilde{u}_2 - \widetilde{u}_1\|_\infty - (3.0002 + 2.849) \geq (\widetilde{u}_1 - \widetilde{u}_2)(0.5, 0.5) - 5.8492$$

$$\geq (1427 - 293) - 5.8492 = 1128.1508.$$

Thus we can deduce that the two solutions on the lower and on the upper solution branches are different.

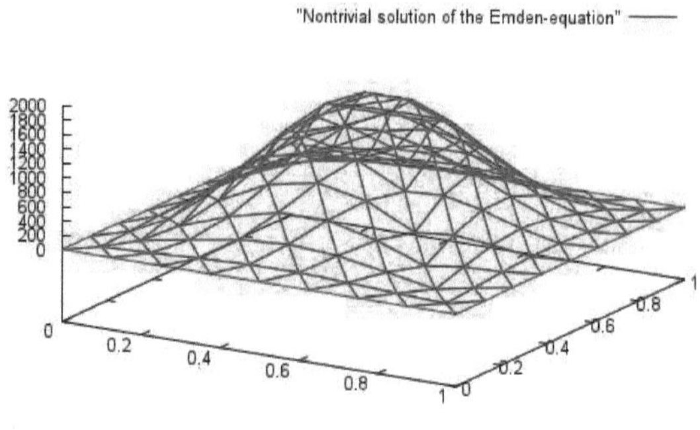

Figure 18: *Nontrivial numerical solution of the Emden-equation on the upper solution branch on the unit square*

(ii) For $\lambda = 0$, which is the true Emden-equation, we computed with 591362 unknowns the following defects:

$$\begin{aligned}
\|\widetilde{\sigma} - \nabla\widetilde{u}\|_{L_2} &\leq 0.0007912\\
\|\widetilde{\sigma} - \nabla\widetilde{u}\|_{L_4} &\leq 0.001192\\
\|rot\ \widetilde{\sigma}\|_{L_2} &\leq 0.010364\\
\|div\ \widetilde{\sigma} + \widetilde{v}\|_{L_2} &\leq 0.019605\\
\|\nabla\widetilde{v} - \widetilde{\rho}\|_{L_2} &\leq 13.6073\\
\|div\ \widetilde{\rho} + F(\widetilde{u})\|_{L_2} &\leq 0.219118\\
\text{and } \|\omega\|_\infty &\leq 1876.54.
\end{aligned}$$

Similar to the case when $\lambda = 200000$ on the upper branch, the smallest eigenvalue λ_1 of the eigenvalue problem (145) with $s = 1$ is smaller than 1, the second smallest eigenvalue λ_2 is larger than 1,

$$\lambda_1 \approx 0.501 \quad \text{and} \quad \lambda_2 \approx 2.761.$$

For λ_1 we used the verified Rayleigh-Ritz upper bound

$$\lambda_1 \leq \overline{\lambda}_1 = 0.5014$$

(see Section 6.3.5). For λ_2 we used again as a rough lower bound a lower bound for the second smallest eigenvalue of the eigenvalue problem (145) with $s = 0$, i.e.,

$$\lambda_2^{(0)} \geq 1.43.$$

This is fortunately still larger then 1. Moreover, despite of the fact that it is "much closer" to 1 as the approximate value for λ_2, we could achieve enclosure for a nontrivial solution of the true Emden-equation with this very rough lower bound.

In this way we get $\widetilde{K} \leq 3.3255$. This gives with $\gamma \leq 0.000024995$ that $K \leq 3.3259$. From the defects we computed

$$\delta = 3.31261,$$

that is small enough to fulfil (82). For the minimal α holds $\alpha_{min} \leq 13.18$. This gives the enclosure of the true solution $u^\star \in H_0^2(\Omega)$

$$\|u^\star - \omega\|_{H_0^2} \leq 13.18,$$

and in particular

$$\|u^\star - \omega\|_\infty \leq 4.81.$$

From $\|\widetilde{u} - \omega\|_\infty \leq 0.0117989$ follows then

$$\|u^\star - \widetilde{u}\|_\infty \leq 4.822.$$

To prove that this solution u^* is different from the trivial solution of the true

Emden-equation, it is enough to show that

$$\|u^*\|_\infty > 0.$$

From the above results we have

$$\|u^*\|_\infty \geq \|\widetilde{u}\|_\infty - \|u^* - \widetilde{u}\|_\infty \geq \widetilde{u}(0.5, 0.5) - 4.822 \geq 1743 - 5.8492 = 1737.1508$$

Thus we can deduce that these two solutions are different. This result is consistent with the statement of Theorem 1.7. Note however, that due to the lack of uniqueness we cannot guarantee the coincidence of the Mountain Pass solution given by Theorem 1.7 with the solution u^* obtained here. Moreover, in the latter case we have not only a pure statement of existence, but furthermore we have some knowledge on the picture of the solution.

2. Disc-like domain $\Omega_\circ$

On the disc-like domain $\Omega_\circ$ the value of the apparent turning point is approximately 41000. The maximum norm of both solution branches was about 300 near to the apparent turning point, see Figure 19.

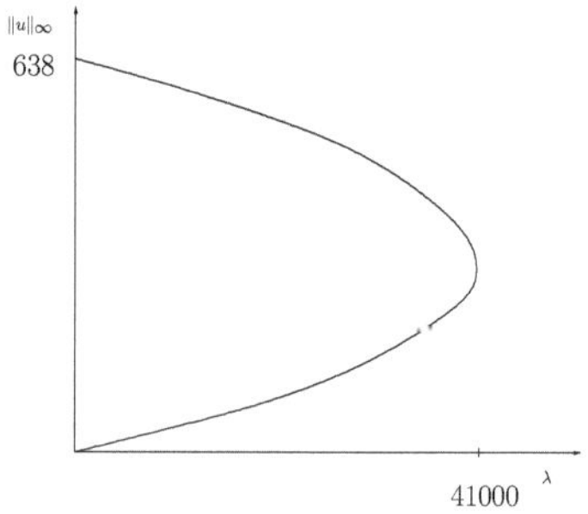

Figure 19: *Maximum norm on the lower and upper solution branches of the Emden-equation on $\Omega_\circ$*

The numerical solutions appear similar to the ones of the Gelfand-equation, see Figures 20 and 21.

We obtained enclosure on the lower solution branch for $\lambda = 5.000$.

We computed the following defects with 9410 unknowns in the finite element approximations:

$$
\begin{aligned}
\|\widetilde{\sigma} - \nabla\widetilde{u}\|_{L_2} &\leq 0.02969 \\
\|\widetilde{\sigma} - \nabla\widetilde{u}\|_{L_4} &\leq 0.37594 \\
\|rot\ \widetilde{\sigma}\|_{L_2} &\leq 0.13117 \\
\|div\ \widetilde{\sigma} + \widetilde{v}\|_{L_2} &\leq 0.40592 \\
\|\nabla\widetilde{v} - \widetilde{\rho}\|_{L_2} &\leq 28.293 \\
\|div\ \widetilde{\rho} + F(\widetilde{u})\|_{L_2} &\leq 57.98 \\
\text{and } \|\omega\|_\infty &\leq 19.0213.
\end{aligned}
$$

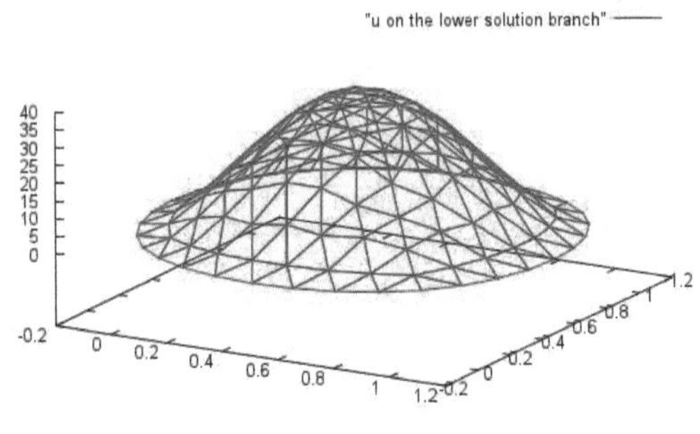

Figure 20: *Numerical solution of the Emden-equation for* $\lambda = 10000$ *on the lower solution branch on* $\Omega_\circ$

To obtain an upper bound for the constant K we used again a rough upper bound for the largest eigenvalue of (67)

$$\kappa_1 \approx 0.065.$$

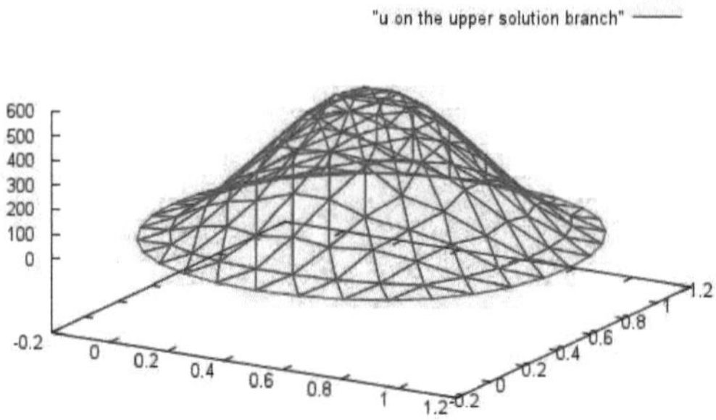

Figure 21: *Numerical solution of the Emden-equation for* $\lambda = 10000$ *on the upper solution branch on* $\Omega_\circ$

To get an upper bound for κ_1 we chose the same comparison problem as in the case of the Gelfand-equation on $\Omega_\circ$, and we obtained

$$\kappa_1 \leq \frac{1}{\lambda_1^{(0)}} \leq 0.108.$$

In this way we get $\widetilde{K} \leq 1.121952$. This gives with $\gamma \leq 0.000858972$ that $K \leq 1.123034$. From the defects we computed

$$\delta = 11.8084,$$

that is small enough to fulfil (82). For the minimal α holds $\alpha_{min} \leq 16.037$. This gives the enclosure of the true solution $u^\star \in H_0^2(\Omega)$

$$\|u^\star - \omega\|_{H_0^2} \leq 16.037,$$

and in particular

$$\|u^\star - \omega\|_\infty \leq 6.0768.$$

From $\|\widetilde{u} - \omega\|_\infty \leq 0.274191$ follows then

$$\|u^\star - \widetilde{u}\|_\infty \leq 6.350991.$$

3. Dumbbell-like domain Ω_d

Also in case of the Emden-equation we found four approximate solution branches, an upper and a lower symmetric branch and two nonsymmetric ones. The two nonsymmetric approximate solutions are reflections of each other with respect to the symmetry-axis $x = 5$ of Ω_d. It can be seen in Figures 25, 26, 27 and 28, how the approximate solutions look like.

The four approximate solution branches behave in many respects similar to the approximate solution branches of the Gelfand-equation. The apparent turning point is approximately 8.3. The nonsymmetric approximate solutions seem to join the symmetric branches in an apparent bifurcation point, close to the apparent turning point.

The two largest approximate eigenvalues of the eigenvalue problem (67) on the upper solution branch are again clustered. For example for $\lambda = 5$ we have

$$\kappa_1 \approx 1.54449, \quad \kappa_2 \approx 1.54443, \quad \kappa_3 \approx 0.35991.$$

Again, the investigation of the simplicity of the turning point requires other methods.

As already mentioned, an essential difference compared to the Gelfand-equation is that the upper and the nonsymmetric approximate solution branches are *bounded near to the origin*, and they tend to an approximate solution of the Emden-equation, i.e., to a solution of the case, when $\lambda = 0$, see Figure 22. Unfortunately we could not achieve enclosure with our method for these nontrivial solutions of the true Emden-equation.

To find these approximate solutions one can use the same technique as we described in Section 4.5, Example 3.

We obtained the following results on Ω_d.

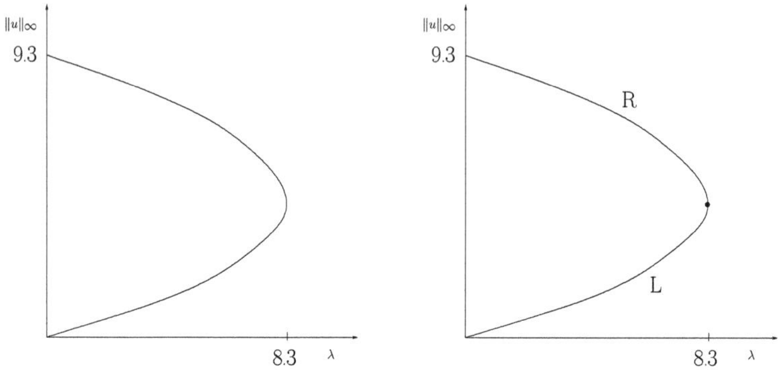

Figure 22: *Maximum norm of the numerical solutions on Ω_d on the symmetric branches and on the non-symmetric branches of the Emden-equation, respectively*

a) The true Emden-equation

The true Emden-equation, i.e., the case when $\lambda = 0$, is particularly important. As already mentioned, we found three nontrivial approximate solutions for this case, but unfortunately we could not achieve enclosure for these solutions. One can see the picture of these solutions in Figures 23 and 24. But these numerical solutions point to the fact that one could show the existence of true nontrivial solutions of the true Emden-equation on Ω_d with more accurate numerics or possibly with other methods.

b) Lower solution branch

For $\lambda = 0.4$ we computed the following approximate values for the defects, where the integrals were approximated by cubature formulas and cubature error terms were neglected. We used 129986 unknowns in the finite element

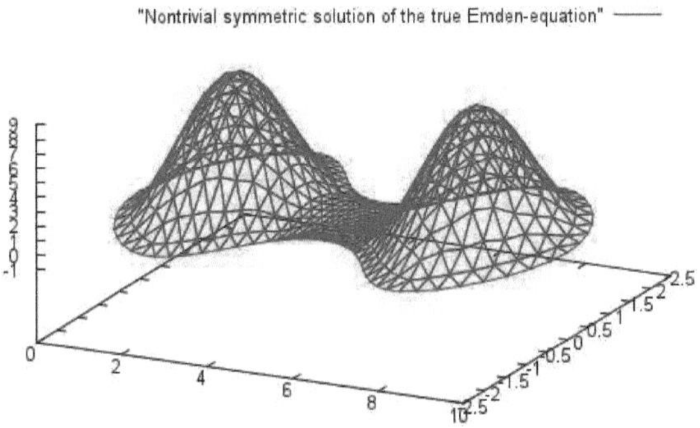

Figure 23: *Symmetric approximate solution of the true Emden-equation on* Ω_d

approximations.

$$
\begin{aligned}
\|\widetilde{\sigma} - \nabla\widetilde{u}\|_{L_2} &\leq 0.0000126164 \\
\|\widetilde{\sigma} - \nabla\widetilde{u}\|_{L_4} &\leq 0.0000167367 \\
\|rot\,\widetilde{\sigma}\|_{L_2} &\leq 0.0000607805 \\
\|div\,\widetilde{\sigma} + \widetilde{v}\|_{L_2} &\leq 0.000156727 \\
\|\nabla\widetilde{v} - \widetilde{\rho}\|_{L_2} &\leq 0.0142451 \\
\|div\,\widetilde{\rho} + F(\widetilde{u})\|_{L_2} &\leq 0.00155363 \\
\text{and } \|\omega\|_\infty &\leq 0.104995.
\end{aligned}
$$

Due to similar arguments as in case of the Gelfand-equation on Ω_d we used the Rayleigh-Ritz approximations to the eigenvalues of (145) to get approximate values for K and α_{min}. Of course, both these neglects (quadrature error and eigenvalue approximation error) spoiled the proof character. But again

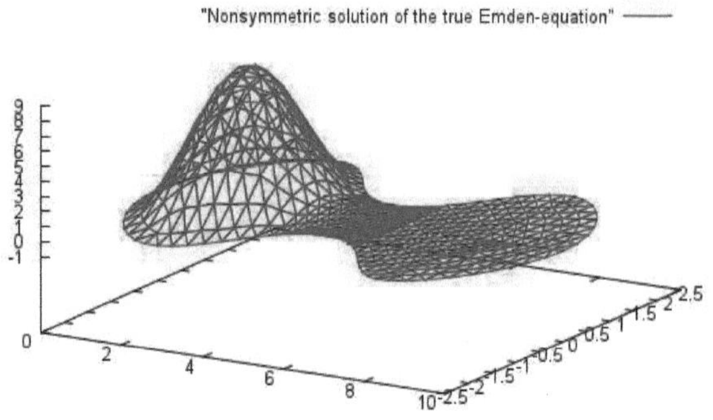

Figure 24: *Nonsymmetric approximate solution of the true Emden-equation on* Ω_d

it indicates that if we will be able to enclose the eigenvalues, then with validated defects the method of Plum will yield us the desired enclosure of the true solution of (73).

From the approximate eigenvalues of (145) with $s = 1$ we obtained 1.187333 as an approximate upper bound for $\widetilde{K}$. With $\gamma \leq 0.0033622$ we get 1.192092 as an approximate upper bound for K.

From the approximate defects we have

$$\delta = 0.0146122,$$

that is small enough to fulfil (78) (with approximate upper bound for K). For the approximate minimal α holds $\alpha_{min} \approx 0.023525$. If we knew the existence of a true solution $u^\star \in H_0^2(\Omega)$ near to the numerical one then taking

$$\|\widetilde{u} - \omega\|_\infty \leq 0.007796$$

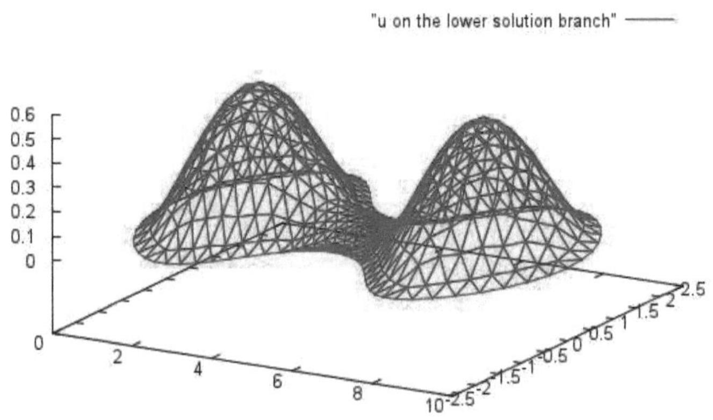

Figure 25: *Numerical solution of the Emden-equation for* $\lambda = 2$ *on the lower approximate solution branch of the Emden-equation on* Ω_d

into account we would get the approximate upper bound 0.04069 for $\|u^\star - \widetilde{u}\|_\infty$.

c) Upper and nonsymmetric solution branches

Due to similar arguments as in the case of the Gelfand-equation we could not achieve enclosure on these branches.

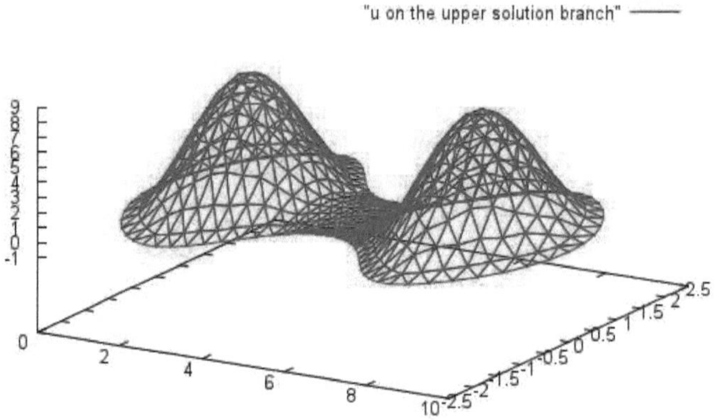

Figure 26: *Numerical solution of the Emden-equation for* $\lambda = 2$ *on the upper approximate solution branch of the Emden-equation on* Ω_d

6 Auxiliary tools

6.1 The div-rot constant

The so-called div-rot constant, that plays an important role in the estimates of ω, is a constant $D > 0$, which fulfils

$$\|div\ \sigma\|_{L_2} \leq D\|rot\ \sigma\|_{L_2} \tag{83}$$

for all $\sigma \in (H_0^1(\Omega))^2$ with $\Delta div\ \sigma = 0$, as we will see in Lemma 6.2. This lemma is a consequence of results in [28]. To be able to prove it, let us summarise these results with the notations of [28].

The Velte Decomposition of the space $\left(H_0^1(\Omega)\right)^2$ is the following orthogonal

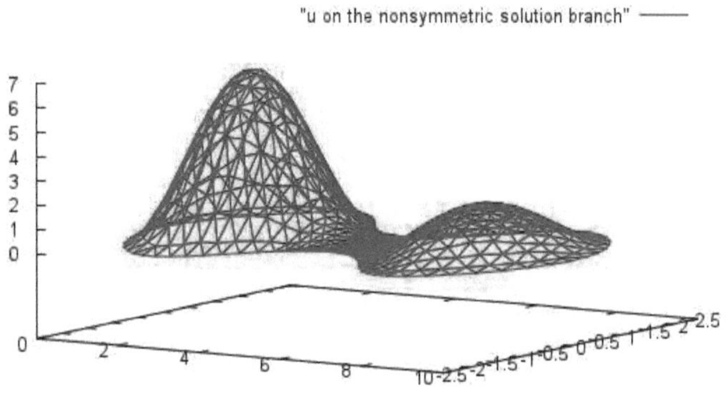

Figure 27: *Numerical solution of the Emden-equation for* $\lambda = 6$ *on the non-symmetric approximate solution branch on* Ω_d

decomposition:

$$\left(H_0^1(\Omega)\right)^2 = V_0 \oplus V_1 \oplus V_\beta, \tag{84}$$

where

$$\begin{aligned} V_0 = \ker div \ &= \left\{\varphi \in \left(H_0^1(\Omega)\right)^2 : div \ \ \varphi = 0\right\}, \\ V_1 = \ker rot \ &= \left\{\varphi \in \left(H_0^1(\Omega)\right)^2 : rot \ \ \varphi = 0\right\}. \end{aligned}$$

To achieve an analogous decomposition of the space $L_2(\Omega)$, we make the following considerations: the operator *div* maps V_0 onto 0, $V_1 \oplus V_\beta$ onto

$$L_{2,0}(\Omega) = \left\{u \in L_2(\Omega) \colon \int_\Omega u \, dx = 0\right\}$$

as an isomorphism (see [10], Lemma 3.2). Moreover, we can write the scalar

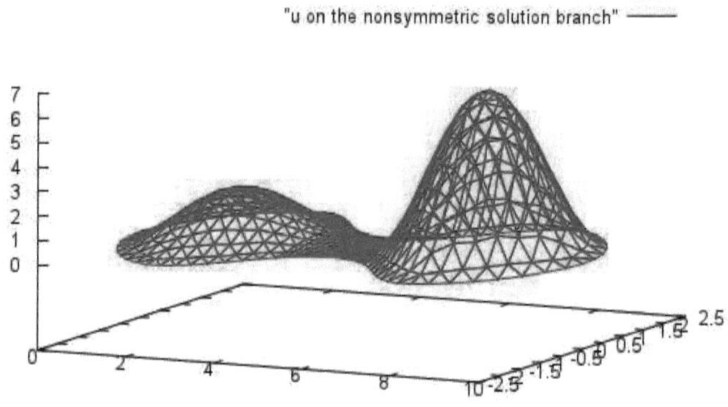

Figure 28: *Numerical solution of the Emden-equation for* $\lambda = 6$ *on the non-symmetric approximate solution branch on* Ω_d

product in $\left(H_0^1(\Omega)\right)^2$ as

$$\langle v, w\rangle_{H_0^1} = \langle \text{div } v, \text{div } w\rangle_{L_2} + \langle \text{rot } v, \text{rot } w\rangle_{L_2} \qquad \text{for } v, w \in \left(H_0^1(\Omega)\right)^2. \quad (85)$$

(For the proof of (85) we choose first $w \in C_0^\infty(\Omega)$, then we get the assertion by partial integration. Using that $C_0^\infty(\Omega)$ is dense in $\left(H_0^1(\Omega)\right)^2$ we obtain (85) for all $v, w \in \left(H_0^1(\Omega)\right)^2$.)

According to (85) we have for $v \in V_1$, $w \in V_\beta$ that

$$0 = \langle v, w\rangle_{H_0^1} = \langle \text{div } v, \text{div } w\rangle_{L_2}.$$

Thus

$$L_{2,0}(\Omega) = div\,(V_1 + V_\beta) = div\,(V_1) \oplus div\,(V_\beta)\,.$$

Therefore

$$L_2(\Omega) = P_0 \oplus P_1 \oplus P_\beta,$$

where P_0 is the space of constant functions, $P_1 = div\ (V_1)$ and $P_\beta = div\ (V_\beta)$.

6.1 Lemma (Stoyan) *(i) P_β is the space of all harmonic functions with zero mean over Ω,*
(ii) $P_1 = rot\ (V_0) \perp rot\ (V_\beta) = P_\beta$,
(iii) In V_β there exists a constant $0 < D$ such that

$$D = \sup_{0 \neq v \in V_\beta} \frac{\|div\ v\|_{L_2}}{\|rot\ v\|_{L_2}}. \tag{86}$$

We remark, that D depends only on the shape of the domain Ω, and not even on its size.

Relation (86) holds not only in the space V_β, but in a larger subset of $(H_0^1(\Omega))^2$. This is the statement of our Lemma.

6.2 Lemma and Definition *With the above constant D of (86) it holds that*

$$\|div\ \sigma\|_{L_2} \leq D\|rot\ \sigma\|_{L_2} \tag{87}$$

for all $\sigma \in (H_0^1(\Omega))^2$ with $\Delta div\ \sigma = 0$. We will call D the div-rot constant.

Proof: From (84) and (i) of Lemma 6.1 follows for $\sigma \in (H_0^1(\Omega))^2$, that $\Delta div\ \sigma = 0$ if and only if $\sigma \in V_\beta + V_0$.

Let $\sigma = v_\beta + v_0$ with $v_\beta \in V_\beta$, $v_0 \in V_0$. Then

$$\|div\ \sigma\|_{L_2} = \|div\ v_\beta\|_{L_2} \overset{\text{Lemma 6.1, (iii)}}{\leq} D\|rot\ v_\beta\|_{L_2}$$

$$\leq D\big(\|rot\ v_\beta\|_{L_2}^2 + \|rot\ v_0\|_{L_2}^2\big)^{\frac{1}{2}} \overset{\text{Lemma 6.1, (ii)}}{=} D\|rot\ (v_\beta + v_0)\|_{L_2} = D\|rot\ \sigma\|_{L_2}.$$

Thus the assertion holds. □

6.3 Remark *The problem of the div-rot inequality (87) is equivalent amongst others to Friedrichs' inequality, i.e., to the following problem: find a constant* Γ *such that for all* $f, g : \Omega \to \mathbb{R}$ *harmonic conjugate functions in* $L_2(\Omega)$ *(i.e.,* $f + ig$ *is a holomorphic function) with the normalisation* $\int_\Omega f\,dx = 0$ *the inequality*

$$\int_\Omega f(x)^2\,dx \le \Gamma \int_\Omega g(x)^2\,dx \tag{88}$$

holds. For the proof see [14] and [9]. The specific connection between the two constants D *and* Γ *is* $\Gamma = D^2$.

6.1.1 Computation of the div-rot constant on star-shaped domains

We showed in the previous section the existence of the div-rot constant. As almost all quantities in this work, this constant also needs to be computed, or at least we need a computable upper bound for it. In the paper [14], Section 6., Horgan and Payne give an upper bound for the constant of Friedrichs' inequality, and thus also for D, in case of a star-shaped C^1-domain Ω.

Let us introduce the notations of the paper. Let Ω be a star-shaped C^1-domain. Let us assume without loss of generality that the star-center is the origin. Let the boundary $\partial\Omega$ be represented in polar coordinates by

$$f : [0, 2\pi] \to \mathbb{R}_+, \quad f(\theta) = r.$$

6.4 Theorem *Let* $h, g : \Omega \to \mathbb{C}$ *be harmonic conjugate functions with the normalisation* $h(0,0) = 0$. *Let* $0 < \alpha < 1$ *be arbitrary and let us denote* $r_M = \max_{\theta\in[0,2\pi]} f(\theta)$. *Then with*

$$P(\alpha,\theta) = \frac{r_M^2}{f^4(\theta)}\left(\frac{-\alpha f^4(\theta) + r_M^2(f^2(\theta) + f'^2(\theta))}{-\alpha^2 f^2(\theta) + \alpha r_M^2}\right) \tag{89}$$

and

$$\widetilde{\Gamma} = \max_{\theta \in [0,2\pi]} P(\alpha, \theta) \tag{90}$$

holds that

$$\int_\Omega h^2 \, dx \leq \widetilde{\Gamma} \int_\Omega g^2 \, dx.$$

6.5 Remark *(i) Observe the different normalisation of the conjugate harmonic function h in Remark* 6.3 *and Theorem* 6.4. *It is easy to see, that* $\Gamma \leq \widetilde{\Gamma}$ *for* Γ *of (88).*

(ii) As all the quantities defining $\widetilde{\Gamma}$ *are directly computable data of the domain, an upper bound for* $\widetilde{\Gamma}$ *is computable.*

Proof of Theorem 6.4 Let the functions H and G be defined via

$$H(x,y) = h^2(x,y) - g^2(x,y) = \mathrm{Re}\left[(h+ig)^2\right]$$

and

$$G(x,y) = 2 \cdot h(x,y) \cdot g(x,y) = \mathrm{Im}\left[(h+ig)^2\right].$$

Thus H and G are harmonic conjugate functions with

$$H(0,0) \leq G(0,0) = 0. \tag{91}$$

Then in polar coordinates using the Cauchy-Riemann equations

$$H(r,\theta) - H(0,\theta) = \int_0^r \frac{\partial H}{\partial \rho}(\rho,\theta)\, d\rho = \int_0^r \frac{1}{\rho}\frac{\partial G}{\partial \theta}(\rho,\theta)\, d\rho.$$

Thus because of (91)

$$H(r,\theta) \leq \int_0^r \frac{1}{\rho}\frac{\partial G}{\partial \theta}(\rho,\theta)\, d\rho. \tag{92}$$

Using (92) we obtain the following estimates:

$$\int_\Omega \frac{H(r,\theta)}{f^2(\theta)}\, d(x,y) \le \int_\Omega \frac{1}{f^2(\theta)} \left[\int_0^r \frac{1}{\rho}\frac{\partial G}{\partial \theta}(\rho,\theta)\, d\rho \right] d(x,y) \tag{93}$$

$$= \int_0^{2\pi}\int_0^{f(\theta)}\int_0^r \frac{r}{\rho f^2(\theta)} \frac{\partial G}{\partial \theta}(\rho,\theta)\, d\rho\, dr\, d\theta$$

$$= \int_0^{2\pi}\int_0^{f(\theta)}\int_\rho^{f(\theta)} \frac{r}{\rho f^2(\theta)} \frac{\partial G}{\partial \theta}(\rho,\theta)\, dr\, d\rho\, d\theta$$

$$= \int_0^{2\pi}\int_0^{f(\theta)} \frac{f^2(\theta)-\rho^2}{2\rho f^2(\theta)} \frac{\partial G}{\partial \theta}(\rho,\theta)\, d\rho\, d\theta$$

$$= \int_\Omega \frac{f^2(\theta)-\rho^2}{2\rho^2 f^2(\theta)} \left(-y\cdot \frac{\partial G}{\partial x}(x,y) + x\cdot \frac{\partial G}{\partial y}(x,y)\right) d(x,y).$$

Let us denote $\psi(\rho,\theta) = \frac{f^2(\theta)-\rho^2}{2\rho^2 f^2(\theta)}$. Then further by integration by parts, using that $\psi(x,y)=0$ on the boundary and that

$$-y\cdot \frac{\partial}{\partial x}\psi(x,y) + x\cdot \frac{\partial}{\partial y}\psi(x,y) = \frac{\partial}{\partial \theta}\psi(r,\theta) = \frac{f'(\theta)}{f^3(\theta)},$$

it holds that

$$\int_\Omega \frac{H(x,y)}{f^2(\theta)}\, d(x,y) \le \int_{\partial\Omega} \psi(x,y)\big(-y\cdot G(x,y)\cdot \nu_1 + x\cdot G(x,y)\cdot \nu_2\big)\, d\sigma$$

$$+ \int_\Omega G(x,y)\Big(\frac{\partial}{\partial x}(y\cdot\psi(x,y)) - \frac{\partial}{\partial y}(x\cdot \psi(x,y))\Big)\, d(x,y)$$

$$= -\int_\Omega G(x,y)\frac{f'(\theta)}{f^3(\theta)}\, d(x,y),$$

where $\nu = (\nu_1,\nu_2)$ denotes the outward normal unit field to $\partial\Omega$. Let us introduce $Q(x,y) = \left|\frac{f'(\theta)}{f(\theta)}\right|$ and substitute the definition of H and G in the above inequality, then we get

$$\int_\Omega \frac{h^2(x,y)}{f^2(\theta)}\, d(x,y) \leq$$

$$\leq \int_\Omega \frac{g^2(x,y)}{f^2(\theta)}\, d(x,y) + 2\int_\Omega \frac{|h(x,y)|\cdot|g(x,y)|\cdot Q(x,y)}{f^2(\theta)}\, d(x,y). \qquad (94)$$

With the arithmetic-geometric mean inequality we get

$$2\cdot|h|\cdot|g|\cdot Q \leq (1-\beta^2)h^2 + Q^2(1-\beta^2)^{-1}g^2, \qquad (95)$$

where $\beta^2 = \alpha\frac{f^2(\theta)}{r_M^2}$ with an arbitrary $0 < \alpha < 1$. Then $1-\beta^2 > 0$ holds. Substituting (95) into (94) yields

$$\int_\Omega h^2(x,y)\, d(x,y)$$

$$\leq \max_{\theta\in[0,2\pi]}\left[\frac{r_M^2}{\alpha f^4(\theta)}\left(\frac{-\alpha f^4(\theta) + r_M^2(f^2(\theta)+f'^2(\theta))}{-\alpha f^2(\theta)+r_M^2}\right)\right]\int_\Omega g^2(x,y)\, d(x,y).$$

Thus the proof is complete. □

For a given domain Ω we can determine $\max_{\theta\in[0,2\pi]} P(\alpha,\theta)$ for all $\alpha\in(0,1)$ and then calculate the infimum over all α.

The following lemma gives us another formula for calculating an upper bound for the div-rot-constant. Although this formula gives a rougher bound, it is often easier to compute. Moreover, if the domain is not too longish, then the ratio $\frac{r_M}{r_m}$ between the maximal and minimal radius of the domain is near to 1. Then formula (96) gives a good upper bound, as for example in case of the disc-like domain $\Omega_\circ$, see Figure 5.

6.6 Lemma *Let us denote* $r_m = \min_{\theta\in[0,2\pi]} f(\theta)$ *and* $K = \max_{\theta\in[0,2\pi]}\left|\frac{f'(\theta)}{f(\theta)}\right|$. *Then with*

$$\widetilde{\Gamma}_1 = \frac{r_M^2}{r_m^2}(K+\sqrt{K^2+1})^2 \qquad (96)$$

holds that $\widetilde{\Gamma} \leq \widetilde{\Gamma}_1$.

Proof: Let $\beta > K$. Then the arithmetic-geometric mean inequality yields

$$2|h||g| \leq \frac{1}{\beta}h^2 + \beta g^2.$$

Using the bounds r_m, r_M and K in (94) we get

$$\int_\Omega h^2\, dx \leq \frac{r_M^2}{r_m^2}\frac{1+K\beta}{1-\frac{K}{\beta}}\int_\Omega g^2\, dx.$$

The infimum of the function $\beta \to \frac{1+K\beta}{1-\frac{K}{\beta}}$ on $]K,\infty[$ is attained at $\beta_0 = K + \sqrt{K^2+1}$ and its value is just $(K + \sqrt{K^2+1})^2$. As the function $x \to (\sqrt{1+x^2}+x)$ is strictly monotone increasing on $\mathbb{R}$, our statement holds. □

If the boundary of Ω is parametrised by a differentiable function

$$g = (g_1, g_2)\colon [a,b] \to \mathbb{R}^2,$$

where $[a,b]$ is a real interval, then one can get the terms $f(\theta)$ and $f'(\theta)$ in the following way: let $h\colon [a,b] \to [0,2\pi]$ denote the angle $h(t) = \theta$ and $k\colon [a,b] \to \mathbb{R}_+$ denote the radius, $k(t) = r$, i.e.,

$$k(t) = \sqrt{g_1(t)^2 + g_2(t)^2},$$

and

$$h(t) = \begin{cases} \arctan\left(\frac{g_2(t)}{g_1(t)}\right) & \text{if } g_1(t) > 0, g_2(t) > 0,\\ \arctan\left(\frac{g_2(t)}{g_1(t)}\right) + \pi & \text{if } g_1(t) < 0,\\ \arctan\left(\frac{g_2(t)}{g_1(t)}\right) + 2\pi & \text{if } g_1(t) > 0, g_2(t) < 0,\\ \frac{\pi}{2} & \text{if } g_1(t) = 0, g_2(t) > 0,\\ \frac{3\pi}{2} & \text{if } g_1(t) = 0, g_2(t) < 0. \end{cases}$$

Then due to $f(\theta) = (k \circ h^{-1})(\theta) = k(t)$ we have

$$f(\theta) = \sqrt{g_1(t)^2 + g_2(t)^2} \quad \text{and} \quad f'(\theta) = \frac{k'(t)}{h'(t)} = k(t)\frac{g_1'(t)g_1(t) + g_2'(t)g_2(t)}{g_2'(t)g_1(t) - g_1'(t)g_2(t)}. \tag{97}$$

6.7 Remark *In [14], Section 6 as a consequence of Theorem 6.4 one can find another formula similar to (96), but without the term* $\left(\frac{r_M}{r_m}\right)^2$. *However, the proof contains some mistakes. Also our numeric calculations for the dumbbell-like domain* Ω_d *points to the fact, that this formula cannot be achieved by maximising* $P(\alpha, \theta)$ *for fixed* α *on the boundary and then minimising the maxima in* α*, as done in the paper. Namely, we can calculate that*

$$\max_{\theta \in [0,2\pi]} \frac{f'(\theta)}{f(\theta)} \leq 15.84,$$

and this means that

$$\max_{\theta \in [0,2\pi]} \left(\left[1 + \left(\frac{f'(\theta)}{f(\theta)} \right)^2 \right]^{\frac{1}{2}} + \frac{|f'(\theta)|}{f(\theta)} \right)^2 \leq 1005.6$$

But if we use plots (e.g. Maple) to get a picture of the function $P(\alpha, \theta)$*, then we find that for some values of* θ *the function* $P(\alpha, \theta)$ *is much larger then the above value* 1005.6 *for all* $\alpha \in]0, 1[$. *We verified as an example that* $P(\alpha, 0.838) \geq 7650$ *for all* $\alpha \in [0, 1]$*, therefore*

$$\min_{\alpha \in (0,1]} \max_{\theta \in [0,2\pi]} P(\alpha, \theta) \geq 7650.$$

Examples:

1. *Square:* An upper bound for the div-rot constant on the unit square is $D = \sqrt{2} + 1$, see [28].

2. *The disc-like domain* $\Omega_\circ$*:* Since this domain is similar to a circular disc

we can apply Lemma 6.6 and formula (97). With the help of a mathematical software such as for example Maple or Mathematica one can easily obtain, that $\max_{\theta\in[0,2\pi]} f'(\theta)/f(\theta) \leq 0.05549$, and that $\frac{r_M}{r_m} = \frac{16}{\sqrt{211}}$. This gives $D \leq 1.089$.

3. *The dumbbell-like domain* Ω_d*:* Since in this case $r_M = 5$ and $r_m = \frac{55}{76}$, the ratio $\frac{r_M}{r_m} \approx 6.91$ is too big to apply Lemma 6.6. We therefore calculate with interval arithmetics $\max_{\theta\in[0,2\pi]} P(\alpha,\theta)$ from (89) with the help of the formula (97) for $\alpha \in (0,1)$. We observe, that this expression is decreasing in α, and we get the lowest upper bound for $\widetilde{\Gamma}$ when α is near to 1. Thus for $\alpha = 0.99999$ we obtain $\widetilde{\Gamma} \leq 7662$, hence the div-rot constant $D \leq 87.6$.

6.2 Imbedding constants

To almost all of our computations imbedding constants ($C_{H_0^1 \hookrightarrow L_2}$, $C_{H_0^2 \hookrightarrow L_2}$, $C_{H_0^1 \hookrightarrow L_4}$, $C_{H_0^1 \hookrightarrow L_6}$, $C_{H_0^2 \hookrightarrow H_0^1}$ and $C_{H^{-1} \hookrightarrow H^{-2}}$) are needed. In this section we explain shortly a possibility to get an upper bound for these constants.

(i) Let us consider first the imbedding constant $C_{H_0^1 \hookrightarrow L_2}$. With the help of the smallest Dirichlet eigenvalue of the Laplace operator we can gain an upper bound for $C_{H_0^1 \hookrightarrow L_2}$, as we will see below. The remaining imbedding constants can also be bounded in terms of $C_{H_0^1 \hookrightarrow L_2}$.

Let us consider therefore the eigenvalue problem

$$-\Delta u = \mu u, \quad u \in H_0^1(\Omega). \tag{98}$$

Let us denote the smallest eigenvalue of (98) by μ_1. From the Min-Max-Principle follows that

$$\mu_1 = \min_{U\subset H_0^1(\Omega),\dim U=1} \max_{u\in U, u\neq 0} \frac{\|\nabla u\|_{L_2}^2}{\|u\|_{L_2}^2} = \min_{u\in H_0^1(\Omega)\setminus\{0\}} \frac{\|\nabla u\|_{L_2}^2}{\|u\|_{L_2}^2}.$$

Therefore

$$\|u\|_{L_2} \le \frac{1}{\sqrt{\mu_1}}\|\nabla u\|_{L_2} \quad \text{for all } u \in H_0^1(\Omega),$$

with equality when u is an eigenfunction associated with μ_1. This yields $C_{H_0^1 \hookrightarrow L_2} = \frac{1}{\sqrt{\mu_1}}$.

(ii) It holds that $C_{H^{-1} \hookrightarrow H^{-2}} \le C_{H_0^2 \hookrightarrow H_0^1}$. To see it let $f \in H^{-1}(\Omega)$. Using the density of $H_0^2(\Omega)$ in $H_0^1(\Omega)$ we get

$$\begin{aligned}
\|f\|_{H^{-2}} &= \sup_{\varphi \in H_0^2(\Omega), \varphi \neq 0} \frac{|f[\varphi]|}{\|\Delta\varphi\|_{L_2}} = \sup_{\varphi \in H_0^2(\Omega), \varphi \neq 0} \frac{|f[\varphi]|}{\|\nabla\varphi\|_{L_2}} \frac{\|\nabla\varphi\|_{L_2}}{\|\Delta\varphi\|_{L_2}} \\
&\le C_{H_0^2 \hookrightarrow H_0^1} \sup_{\varphi \in H_0^2(\Omega), \varphi \neq 0} \frac{|f[\varphi]|}{\|\nabla\varphi\|_{L_2}} \overset{\text{density}}{=} C_{H_0^2 \hookrightarrow H_0^1} \sup_{\varphi \in H_0^1(\Omega), \varphi \neq 0} \frac{|f[\varphi]|}{\|\nabla\varphi\|_{L_2}} \\
&= C_{H_0^2 \hookrightarrow H_0^1} \|f\|_{H^{-1}}.
\end{aligned}$$

(iii) Analogously to case (i) we can obtain an upper bound for $C_{H_0^2 \hookrightarrow H_0^1}$ with the help of the smallest eigenvalue of the following eigenvalue problem

$$\Delta^2 u = \widetilde{\mu}(-\Delta u), \quad u \in H_0^2(\Omega). \tag{99}$$

Let $\widetilde{\mu}_1$ denote the smallest eigenvalue of (99). From the Min-Max-Principle follows that

$$\widetilde{\mu}_1 = \min_{U \subset H_0^2(\Omega), \dim U = 1} \max_{u \in U, u \neq 0} \frac{\|\Delta u\|_{L_2}^2}{\|\nabla u\|_{L_2}^2} = \min_{u \in H_0^2(\Omega) \setminus \{0\}} \frac{\|\Delta u\|_{L_2}^2}{\|\nabla u\|_{L_2}^2}.$$

Therefore

$$\|\nabla u\|_{L_2} \le \frac{1}{\sqrt{\widetilde{\mu}_1}}\|\Delta u\|_{L_2} \quad \text{for all } u \in H_0^2(\Omega),$$

with equality when u is an eigenfunction of (99) associated with $\widetilde{\mu}_1$. This yields $C_{H_0^2 \hookrightarrow H_0^1} = \frac{1}{\sqrt{\widetilde{\mu}_1}}$.

Another way to get an upper bound for $C_{H_0^2 \hookrightarrow H_0^1}$ is to use the embedding constant $C_{H_0^1 \hookrightarrow L_2}$. Namely, it holds that $C_{H_0^2 \hookrightarrow H_0^1} \leq C_{H_0^1 \hookrightarrow L_2}$. To see it let $u \in H_0^2(\Omega)$. Then with the notation $\sigma = (\sigma_1, \sigma_2) = \nabla u \in (H_0^1(\Omega))^2$ we have $rot\ \sigma = 0$. Using (85) we get

$$\|u\|_{H_0^1}^2 = \|\sigma\|_{L_2}^2 = \|\sigma_1\|_{L_2}^2 + \|\sigma_2\|_{L_2}^2 \leq C_{H_0^1 \hookrightarrow L_2}^2 (\|\nabla \sigma_1\|_{L_2}^2 + \|\nabla \sigma_2\|_{L_2}^2)$$

$$= C_{H_0^1 \hookrightarrow L_2}^2 \|\sigma\|_{H_0^1}^2 = C_{H_0^1 \hookrightarrow L_2}^2 \|div\ \sigma\|_{L_2}^2 = C_{H_0^1 \hookrightarrow L_2}^2 \|\Delta u\|_{L_2}^2 = C_{H_0^1 \hookrightarrow L_2}^2 \|u\|_{H_0^2}^2.$$

(iv) Again, analogously to case (i) we can obtain an upper bound for $C_{H_0^2 \hookrightarrow L_2}$ with the help of the smallest eigenvalue of the biharmonic operator. Thus let $\hat{\mu}_1$ be the smallest eigenvalue of the eigenvalue problem

$$\Delta^2 u = \hat{\mu} u, \quad u \in H_0^2(\Omega). \tag{100}$$

From the Min-Max-Principle follows that

$$\hat{\mu}_1 = \min_{U \subset H_0^2(\Omega), \dim\ U = 1} \max_{u \in U, u \neq 0} \frac{\|\Delta u\|_{L_2}^2}{\|u\|_{L_2}^2} = \min_{u \in H_0^2(\Omega) \setminus \{0\}} \frac{\|\Delta u\|_{L_2}^2}{\|u\|_{L_2}^2}.$$

Therefore

$$\|u\|_{L_2} \leq \frac{1}{\sqrt{\hat{\mu}_1}} \|\Delta u\|_{L_2} \quad \text{for all } u \in H_0^2(\Omega),$$

with equality when u is an eigenfunction associated with $\hat{\mu}_1$. This yields $C_{H_0^2 \hookrightarrow L_2} = \frac{1}{\sqrt{\hat{\mu}_1}}$.

Another possibility to obtain an upper bound for $C_{H_0^2 \hookrightarrow L_2}$ is to use $C_{H_0^2 \hookrightarrow L_2} \leq C_{H_0^2 \hookrightarrow H_0^1} \cdot C_{H_0^1 \hookrightarrow L_2}$ and (ii) to get $C_{H_0^2 \hookrightarrow L_2} \leq C_{H_0^1 \hookrightarrow L_2}^2$.

(v) At last it holds that $C_{H_0^1 \hookrightarrow L_4} \leq \left(\frac{1}{2} C_{H_0^1 \hookrightarrow L_2}^2\right)^{1/4}$ and $C_{H_0^1 \hookrightarrow L_6} \leq \left(\frac{3}{4} C_{H_0^1 \hookrightarrow L_2}\right)^{1/3}$. For the proof we refer to [26], Lemma 6.

In case (iii) and (iv) the bounds in terms of $C_{H_0^1 \hookrightarrow L_2}$ give in general rougher upper bounds than the ones containing the smallest eigenvalue of the given eigenvalue problems. But usually it is much easier to compute bounds for the eigenvalues of the second order Laplace operator as to compute bounds for the above eigenvalue problems for the fourth order biharmonic operator. Thus the bounds in terms of $C_{H_0^1 \hookrightarrow L_2}$ are often also useful.

We will describe a method for obtaining bounds for the smallest eigenvalue of the Laplace and the biharmonic operator in the next section.

6.3 Eigenvalue bounds and eigenvalue homotopy

6.3.1 Eigenvalue bounds

In this section we describe a method, how two-sided eigenvalue bounds for problems of the form

$$\langle u, \varphi \rangle_H = \lambda N(u, \varphi), \qquad \text{for all } \varphi \in H, \tag{101}$$

can be obtained, where $(H, \langle \cdot, \cdot \rangle_H)$ denotes a separable Hilbert space and N a bounded, positive definite, symmetric bilinear form on H. Let us denote by σ_0 the infimum of the essential spectrum of problem (101).

The method of Rayleigh-Ritz ensures upper eigenvalue bounds for the n smallest eigenvalues $\lambda_1 \leq \cdots \leq \lambda_n$ of problem (101), provided that at least n eigenvalues of (101) below σ_0 exist.

6.8 Theorem (Method of Rayleigh-Ritz) *Let $n \in \mathbb{N}$ and $u_1, \ldots, u_n \in H$ be linearly independent functions. Let us further define the matrices*

$$A_0 = \left(\langle u_i, u_j \rangle_H \right)_{i,j=1,\ldots,n}, \qquad A_1 = \left(N(u_i, u_j) \right)_{i,j=1,\ldots,n}.$$

Let the eigenvalues of the matrix eigenvalue problem

$$A_0 x = \kappa A_1 x \tag{102}$$

be denoted by $\kappa_1 \leq \cdots \leq \kappa_n$. If $\kappa_n \leq \sigma_0$, then problem (101) *has at least n eigenvalues below σ_0 and it holds that*

$$\lambda_i \leq \kappa_i \qquad i = 1, \ldots, n.$$

For the proof we refer to [22]. Observe, that to apply this method, one only has to enclose eigenvalues of a matrix eigenvalue problem. Since there exist direct methods for that problem, the Rayleigh-Ritz method can be easily realised. If one chooses $u_1, \ldots, u_n$ as approximate eigenfunctions of (101), then it ensures very good upper bounds for the eigenvalues $\lambda_1, \ldots, \lambda_n$. This means in particular, that in this case, the values $\kappa_1, \ldots, \kappa_n$ also serve as very good approximations to $\lambda_1, \ldots, \lambda_n$. One can obtain approximate eigenfunctions by numerical means, e.g. by finite element approximation of (101).

It is a much harder task to obtain lower eigenvalue bounds. One possibility is to use the method of Lehmann-Goerisch. The following version admitting essential spectrum is due to Plum (see e.g. [5]).

6.9 Theorem *Let $(X, b(\cdot,\cdot))$ denote a complex Hilbert space and $T : H \to X$ an isometric linear operator, i.e., $b(T\varphi, T\psi) = \langle \varphi, \psi \rangle_H$ for all $\varphi, \psi \in H$. Let $n \in \mathbb{N}$ and $u_1, \ldots, u_n \in H$ be linearly independent functions. Let $w_1, \ldots, w_n$ satisfy*

$$b(T\varphi, w_i) = N(\varphi, u_i) \qquad \text{for all } \varphi \in H, \quad i = 1, \ldots, n. \tag{103}$$

Let us further define the matrices

$$A_0 = \left(\langle u_i, u_j \rangle_H\right)_{i,j=1,\ldots,n}, \; A_1 = \left(N(u_i, u_j)\right)_{i,j=1,\ldots,n}, \; A_2 = \left(b(w_i, w_j)\right)_{i,j=1,\ldots,n}.$$

Let $0 < \beta \leq \sigma_0$ be given such that
(i) there are at most finitely many eigenvalues of (101) *below β,*

(ii) the matrix

$$A_0 - \beta A_1$$

is negative definite.
Let the (negative) eigenvalues of the matrix eigenvalue problem

$$(A_0 - \beta A_1)x = \kappa(A_0 - 2\beta A_1 + \beta^2 A_2)x \tag{104}$$

be denoted by $\kappa_1 \leq \cdots \leq \kappa_n < 0$. *Then problem* (101) *has at least* n *eigenvalues below* β *and for the* n *largest of them* $\lambda_n \leq \cdots \leq \lambda_1$ *it holds that*

$$\lambda_i \geq \beta - \frac{\beta}{1-\kappa_i}, \qquad i = 1, \ldots, n. \tag{105}$$

For the proof we refer to [5]. (The above version of this theorem is weaker, than the one in [5]. We remark, that from assumption (ii) of Theorem 6.9 follows assumption (42) in Theorem 3 in [5]. This is the special case, when the matrix eigenvalue problem (104) has the full number, i.e., $k = n$ negative eigenvalues.)

We remark further, that the matrix eigenvalue problem (104) has indeed n negative eigenvalues, since the matrix $A_0 - 2\beta A_1 + \beta^2 A_2$ is positive definite. To see it, let $x = (x_1, \ldots, x_n) \in \mathbb{R}^n$, $u = \sum_{i=1}^n x_i u_i$, $w = \sum_{i=1}^n x_i w_i$. Then

$$x^T(A_0 - 2\beta A_1 + \beta^2 A_2)x = \langle u, u\rangle_H - 2\beta N(u,u) + \beta^2 b(w,w)$$

$$= b(Tu, Tu) - 2\beta b(Tu, w) + \beta^2 b(w,w) = b(Tu - \beta w, Tu - \beta w) \geq 0.$$

Moreover, if equality holds here, i.e. if $Tu = \beta w$, then

$$x^T A_0 x = \langle u, u\rangle_H = b(Tu, Tu) = \beta b(Tu, w) \overset{(103)}{=} \beta N(u,u) = \beta x^T A_1 x.$$

Therefore

$$x^T(A_0 - \beta A_1)x = 0,$$

which means $x = 0$, since due to assumption (ii) of Theorem 6.9 the matrix $A_0 - \beta A_1$ is negative definite.

Similarly, as by the Rayleigh-Ritz Theorem, we choose the functions $u_1,\dots,u_n$ as approximate eigenfunctions. But how shall we choose the functions w_1, $\dots$, w_n? If $u_1, \dots, u_n$ were the exact eigenfunctions, and $w_i = \frac{1}{\lambda_i} T u_i$ $(i = 1, \dots, n)$, then equality (103) would clearly be satisfied (this corresponds to the original Lehmann method, see [15]). Thus, if $u_1, \dots, u_n$ are approximate eigenfunctions, then one should choose $w_i \approx \frac{1}{\lambda_i} T u_i$ $(i = 1, \dots, n)$, of course with (103) as a hard side condition.

The next question is, how we can use the Lehmann-Goerisch method, combined with the Rayleigh-Ritz method, to obtain enclosure for $\lambda_1, \dots, \lambda_n$. For that purpose, we can proceed as follows. First we compute approximate eigenelements $u_1, \dots, u_n \in H$. We can apply the Rayleigh-Ritz method, which yields us the upper bounds $\overline{\lambda}_1, \dots, \overline{\lambda}_n$ for $\lambda_1, \dots, \lambda_n$.

Then we apply Lehmann-Goerisch method: First we define suitably the space $(X, b(\cdot, \cdot))$, the isometry T and the functions $w_1, \dots, w_n$ (depending on the given problem). Then we choose a $\beta \leq \sigma_0$ such that

$$\overline{\lambda}_n < \beta \leq \lambda_{n+1}. \tag{106}$$

Assumptions (i) and (ii) follow from (106). Then Theorem 6.9 yields lower bounds $\underline{\lambda}_1, \dots, \underline{\lambda}_n$ for $\lambda_1, \dots, \lambda_n$. Therefore the intervals $[\underline{\lambda}_i, \overline{\lambda}_i]$ are enclosing intervals for λ_i for $i = 1, \dots, n$.

We show a possible choice for X, T, b and $w_1, \dots, w_n$ for two different problems in Sections 6.3.4 and 6.3.5, respectively. A method for finding β is described in Sections 6.3.2 and 6.3.3.

In the special case $n = 1$ of Theorem 6.9, we can simplify assumption (ii) and we can calculate the lower bound in (105) directly. This is the statement of the following corollary.

6.10 Corollary *Let $(X, b(\cdot,\cdot))$ denote a complex Hilbert space and $T : H \to X$ an isometric linear operator. Let $u \in H$ and let w satisfy*

$$b(T\varphi, w) = N(\varphi, u) \qquad \text{for all } \varphi \in H. \tag{107}$$

Let $0 < \beta \leq \sigma_0$ be given such that
(i) there are at most finitely many eigenvalues of (101) below β,
(ii) and

$$\frac{\langle u, u\rangle_H}{N(u,u)} < \beta. \tag{108}$$

Then there exists an eigenvalue κ of problem (101), which satisfies

$$\frac{\beta N(u,u) - \langle u, u\rangle_H}{\beta b(w,w) - N(u,u)} \leq \kappa < \beta. \tag{109}$$

6.11 Remark *Observe, that if u_n is the exact eigenfunction to the eigenvalue λ_n and $w = \frac{1}{\lambda_n} T u_n$, then*

$$b(w,w) = \frac{1}{\lambda_n^2} b(Tu_n, Tu_n) = \frac{1}{\lambda_n^2}\langle u_n, u_n\rangle_H,$$

and thus

$$\frac{\beta N(u_n,u_n) - \langle u_n, u_n\rangle_H}{\beta b(w,w) - N(u_n,u_n)} = \frac{\frac{\beta}{\lambda_n}\langle u_n, u_n\rangle_H - \langle u_n, u_n\rangle_H}{\frac{\beta}{\lambda_n^2}\langle u_n, u_n\rangle_H - \frac{1}{\lambda_n}\langle u_n, u_n\rangle_H} = \frac{\beta\lambda_n - \lambda_n^2}{\beta - \lambda_n} = \lambda_n.$$

Thus, if u is approximate eigenfunction to the eigenvalue λ_n, then (109) gives a good lower bound for λ_n.

6.3.2 Comparison problems

For the method of Lehmann-Goerisch a constant β satisfying (106) is required. One can obtain a suitable constant in many cases with the help of a so-called comparison problem. Roughly speaking a comparison problem is another eigenvalue problem such that its eigenvalues are index-wise smaller than the eigenvalues of the original problem (101). More precisely, let $(H_0, \langle\cdot,\cdot\rangle_{H_0})$ be another separable Hilbert space and N_0 a bounded, positive definite, symmetric bilinear form on H_0. We consider the following eigenvalue problem

$$\langle u, \varphi\rangle_{H_0} = \lambda^{(0)} N_0(u, \varphi), \qquad \text{for all } \varphi \in H_0. \tag{110}$$

Let us assume that the infimum of the essential spectrum of problem (110) coincides with σ_0 (the infimum of the essential spectrum of problem (101)). Moreover, to ensure the desired connection between the eigenvalues of problem (110) and of (101) let us assume that

$$H \subset H_0, \quad \text{and} \quad \frac{\langle u, u\rangle_{H_0}}{N_0(u, u)} \leq \frac{\langle u, u\rangle_H}{N(u, u)} \quad \text{for all } u \in H. \tag{111}$$

Provided that at least $n+1$ eigenvalues of (101) below σ_0 exist, the smallest of which we denote again by $\lambda_1, \ldots, \lambda_{n+1}$, we obtain by the Min-Max Principle from (111) that

$$\lambda_i^{(0)} \leq \lambda_i, \quad i = 1, \ldots, n+1, \tag{112}$$

where $\lambda_1^{(0)}, \ldots, \lambda_{n+1}^{(0)}$ denote the $n+1$ smallest eigenvalues of problem (110). Therefore, $\beta = \lambda_{n+1}^{(0)}$ can be chosen in (106) provided that

$$\overline{\lambda_n} < \lambda_{n+1}^{(0)}. \tag{113}$$

This method is only practicable of course, if we have some knowledge on the spectrum of the comparison problem, for example if we know precisely its eigenvalues, or if we can at least easily calculate rough lower bounds for them. Furthermore, assumption (113) is satisfied in general only if the comparison problem is not too "far away" from the original problem.

6.3.3 Eigenvalue homotopy

To obtain a constant β satisfying (106) with the help of a comparison problem is of course not always possible, since in many cases one cannot find an appropriate problem satisfying also (113). There is a more general possibility to gain the desired constant β via the so-called homotopy method. This method is in some sense a generalisation of the method using a comparison problem.

In the course of the homotopy a so-called base-problem is connected to our given problem via a sequence of intermediate problems, such that in each step the previous problem serves as a comparison problem for the next problem. The base problem is chosen such that we have some knowledge on its spectrum and moreover that in each step the eigenvalues of the intermediate problems are increasing. With the help of the knowledge on the spectrum of the base problem we gain information about the eigenvalues of the next problem in the homotopy, and then in the further steps about the eigenvalues of the following problems, until we arrive at our given problem.

More precisely, for all $s \in [0,1]$ let $(H_s, \langle\cdot,\cdot\rangle_{H_s})$ be a separable Hilbert space and N_s a bounded, positive definite, symmetric bilinear form on H_s such that $N_1 = N$ and $H_1 = H$. We consider the family of eigenvalue problem

$$\langle u, \varphi\rangle_{H_s} = \lambda^{(s)} N_s(u,\varphi), \qquad \text{for all } \varphi \in H_s. \tag{114}$$

Let us assume, that the infimum of the essential spectra of problems (114) for all $s \in [0,1]$ coincides with σ_0 (the infimum of the essential spectrum of problem (101), i.e. problem (114) with $s = 1$). To ensure the desired monotonicity property of the eigenvalues let us assume that

$$H_{s_2} \subset H_{s_1}, \qquad \text{and} \qquad \frac{\langle u,u\rangle_{H_{s_1}}}{N_{s_1}(u,u)} \le \frac{\langle u,u\rangle_{H_{s_2}}}{N_{s_2}(u,u)} \qquad \text{for all } s_1 \le s_2,\ u \in H_{s_2}. \tag{115}$$

Again assuming that (at least) $n+1$ eigenvalues of problem (114) below σ_0 exist, we obtain by the Min-Max Principle from (115) that indeed for $s_1 \le s_2$

$$\lambda_i^{(s_1)} \le \lambda_i^{(s_2)}, \quad i = 1, \dots, n+1, \tag{116}$$

where $\lambda_1^{(s)}, \dots, \lambda_{n+1}^{(s)}$ denote the $n+1$ smallest eigenvalues of problem (114).

Moreover, let some $\beta_0 \le \sigma_0$ and $m \in \mathbb{N}$ be given such that the base problem, i.e., problem (114) with $s = 0$, has precisely m eigenvalues in $(0, \beta_0)$. Because of (116) problem (114) has at most m eigenvalues in $(0, \beta_0)$ for all $s \in [0, 1]$.

The homotopy method works as follows: we start with the m-th eigenvalue of the base problem. Let us assume, that the gap between $\lambda_m^{(0)}$ and β_0 is sufficiently large. Then for some $s_1 > 0$ we compute approximate eigenfunctions $v_1^{(s_1)}, \dots, v_m^{(s_1)}$ and approximate eigenvalues $\widetilde{\lambda}_1^{(s_1)} \le \dots \le \widetilde{\lambda}_m^{(s_1)}$ with the Rayleigh-Ritz method. We will use the approximate eigenfunctions for exact computations, while the approximate eigenvalues "only" for getting a conjecture, how the exact eigenvalues behave. Our aim is now to apply Corollary 6.10 to problem

$$\langle u, \varphi \rangle_{H_{s_1}} = \lambda^{(s_1)} N_{s_1}(u, \varphi), \qquad \text{for all } \varphi \in H_{s_1}, \tag{117}$$

with $u = v_m^{(s_1)}$ and $\beta = \beta_0$. Since problem (114) has at most m eigenvalues in $(0, \beta_0)$, condition (i) is satisfied for all $s \in [0, 1]$. Then we only have to check if

$$\frac{\langle v_m^{(s_1)}, v_m^{(s_1)} \rangle_{H_{s_1}}}{N_{s_1}(v_m^{(s_1)}, v_m^{(s_1)})} < \beta_0. \tag{118}$$

In the affirmative case Corollary 6.10 yields the existence of an eigenvalue κ with

$$\beta_1 := \frac{\beta_0 N_{s_1}(v_m^{(s_1)}, v_m^{(s_1)}) - \langle v_m^{(s_1)}, v_m^{(s_1)} \rangle_{H_{s_1}}}{\beta_0 b_{s_1}(w_{s_1}, w_{s_1}) - N_{s_1}(v_m^{(s_1)}, v_m^{(s_1)})} \le \kappa < \beta_0.$$

Since the base problem has precisely m eigenvalues in $(0, \beta_0)$, due to (116), problem (117) has at most $m-1$ eigenvalues in $(0, \beta_1)$.

In general, there exist infinitely many s_1, which satisfy inequality (118). We choose from the suitable values s_1 as large as possible. Then consequently $\beta_0 - \beta_1$ is small. If the approximations $\widetilde{\lambda}_{m-1}^{(s_1)}$ and $\widetilde{\lambda}_m^{(s_1)}$ are well separated, (and thus - we hope - also $\lambda_{m-1}^{(s_1)}$ and $\lambda_m^{(s_1)}$, if the computations are precise enough), then we can expect, that the only eigenvalue in $[\beta_1, \beta_0)$ is $\lambda_m^{(s_1)}$. Then problem (117) has precisely $m-1$ eigenvalues in $(0, \beta_1)$. This last statement is of course not proved, but it is not necessary. We continue our homotopy, and we verify the analogous statement at the last homotopy step with the help of the Rayleigh-Ritz upper bounds, which also prove our expectations in the intermediate steps. (We could prove this expectation by using the Rayleigh-Ritz upper bounds, i.e., we could check, if $\overline{\lambda}_{m-1}^{(s_1)} < \beta_1$. But if the computations are not precise enough, it can occur that $\lambda_{m-1}^{(s_1)} < \beta_1 < \overline{\lambda}_{m-1}^{(s_1)}$.)

Then we come to the second homotopy step: we proceed as in the first step, now with $m-1$ in place of m, β_1 in place of β_0 and s_1 in place of 0. We obtain then $m-2$, β_2 and s_2.

The algorithm comes to an end in l steps, if either $s_l = 1$ and $m - l \geq 0$ or $s_l < 1$ and $m = l$.

In the first case, if $s_l = 1$, then we "arrived" at our original problem (101) and we can deduce that it has at most $m - l$ eigenvalues in $(0, \beta_l)$. If $m = l$, then β_l yields a lower bound for the first eigenvalue of (101). Otherwise we use the Rayleigh-Ritz upper bounds to verify our expectation: if $\overline{\lambda}_{m-l}^{(1)} < \beta_l$ holds, then problem (101) has indeed precisely $m - l$ eigenvalues in $(0, \beta_l)$ and we found our β satisfying

$$\overline{\lambda}_n < \beta \leq \lambda_{n+1}$$

with $n = m - l$ and $\beta = \beta_l$.

In the second case, if $s_l < 1$ and $m = l$ then problem

$$\langle u, \varphi \rangle_{H_{s_l}} = \lambda^{(s_l)} N_{s_l}(u, \varphi), \qquad \text{for all } \varphi \in H_{s_l} \tag{119}$$

has at most $m - l = 0$ eigenvalues in $(0, \beta_l)$. Because of (116) also problem (101) has no eigenvalues in $(0, \beta_l)$. Thus in this case we obtained a lower bound for the first eigenvalue of (101), i.e., $\beta_l \le \lambda_1$.

In all these steps we assumed, that the gap between $\lambda_m^{(0)}$ and β_0, and $\lambda_{m-1}^{(s_1)}$ and $\lambda_m^{(s_1)}$...etc. is sufficiently large. If in a step this assumption is not fulfilled, i.e., that the eigenvalues are clustered, then one has to apply Theorem 6.9 instead of Corollary 6.10 with n equals to the number of elements in the cluster. Then one has to "drop" the whole cluster and correspondingly reduce the number of eigenvalues by k instead of 1.

We consider the following two more special cases of the general homotopy method.

a) Linear homotopy

In some applications we can choose $H_s = H$ (with equal inner products) for all $s \in [0, 1]$ and, with N_0 given such that

$$N_1(u, u) \le N_0(u, u) \qquad \text{for all } u \in H_1, \tag{120}$$

choose the linear homotopy

$$N_s = (1 - s)N_0 + sN_1.$$

Due to (120) assumption (115) is satisfied. Thus the above homotopy method can be applied.

We will see an example for linear homotopy at the end of Section 6.3.5, where the nonconstant coefficients of the given elliptic eigenvalue problem are homotopically connected with constant coefficients (coefficient homotopy).

b) Domain homotopy

We demonstrate the general homotopy for the following eigenvalue problem

$$u \in H_0^2(\Omega), \quad \Delta^2 u = \lambda u,$$

where $\Omega \subset \mathbb{R}^n$ is a given domain. If we do not have enough information on the eigenvalues on Ω, but we have knowledge on the eigenvalues of the same eigenvalue problem on some larger domain $\Omega_0 \supset \Omega$ (e.g. on a rectangular domain), then we may connect Ω_0 with $\Omega = \Omega_1$ via intermediate domains Ω_s for $s \in [0,1]$ such that $\Omega_{s_2} \subset \Omega_{s_1}$ for $s_1 < s_2$, and on each domain we consider the "same" eigenvalue problem. In the above abstract setting then

$$H_0 = H_0^2(\Omega_0), \qquad H_s = \{u \in H_0 \colon u \equiv 0 \text{ on } \Omega_0 \setminus \Omega_s\},$$
$$\text{with} \quad \langle \cdot, \cdot \rangle_{H_s} = \langle \cdot, \cdot \rangle_{H_0} \text{ on } H_s, \tag{121}$$

further

$$N_0(u, \varphi) = \int_{\Omega_0} u\varphi \, dx \quad \text{and} \quad N_s = N_0 \text{ on } H_s$$

for all $s \in [0,1]$.

Clearly assumption (115) is fulfilled, thus the above homotopy method can be applied.

We remark that the Dirichlet boundary conditions are essential in this case, since for example with Neumann boundary conditions the extension by zero in (121) would be in general discontinuous, and thus lead out of the required Sobolev space.

6.3.4 Application to the Laplace problem

As we have seen in Chapter 6.2 the value $\frac{1}{\sqrt{\mu_1}}$ provides an upper bound for the imbedding constant $C_{H_0^1 \hookrightarrow L_2}$, where μ_1 is the smallest eigenvalue of the Laplace operator. Thus we are aiming at a lower bound for the smallest eigenvalue of problem

$$-\Delta u = \mu u, \quad u \in H_0^1(\Omega),$$

i.e.,

$$\int_\Omega \nabla u \nabla \varphi \, dx = \mu \int_\Omega u\varphi \, dx \quad \text{for all } \varphi \in H_0^1(\Omega),$$

with $u \in H_0^1(\Omega)$.

This is equivalent to

$$\int_\Omega \nabla u \nabla \varphi \, dx + \gamma \int_\Omega u\varphi \, dx = \underbrace{(\gamma + \mu)}_{:=\nu} \int_\Omega u\varphi \, dx \quad \text{for all } \varphi \in H_0^1(\Omega), \quad (122)$$

with $\gamma > 0$ and $u \in H_0^1(\Omega)$. We will make clear the role of γ later. We use the method of Rayleigh-Ritz and Lehmann-Goerisch with the following casting using the notations of Section 6.3.1: the Hilbert space $H = H_0^1(\Omega)$ provided with the inner product

$$\langle \varphi, \psi \rangle_{H_0^1,\gamma} = \int_\Omega \nabla \varphi \nabla \psi \, dx + \gamma \int_\Omega \varphi\psi \, dx.$$

The right-hand side is $N(\varphi, \psi) = \int_\Omega \varphi\psi \, dx$.

Let us define the Hilbert space $(X, b(\cdot,\cdot))$ and the isometry $T : H \to X$ as follows:

$$X = \left(L_2(\Omega)\right)^3, \qquad T\varphi = (\nabla \varphi, \varphi) \quad \text{for all } \varphi \in H_0^1(\Omega),$$

$$b(\varphi, \psi) = \langle \varphi_1, \psi_1 \rangle_{L_2} + \langle \varphi_2, \psi_2 \rangle_{L_2} + \gamma \langle \varphi_3, \psi_3 \rangle_{L_2}$$
$$\text{for all } \varphi = (\varphi_1, \varphi_2, \varphi_3), \psi = (\psi_1, \psi_2, \psi_3) \in \left(L_2(\Omega)\right)^3.$$

Then the assumption $b(T\varphi, T\psi) = \langle \varphi, \psi \rangle_{H^1_{0,\gamma}}$ is obviously satisfied.

Let $(\widetilde{u}, \nu) \in H^1_0(\Omega) \times \mathbb{R}$ be an approximate eigenpair to the exact eigenpair (u_e, ν_e). Let us choose now the vector $w = (w_1, w_2, w_3) \in X$ satisfying (107). Condition (107) reads in this context

$$\int_\Omega \nabla\varphi \cdot \chi \, dx + \gamma \int_\Omega \varphi w_3 \, dx = \int_\Omega \varphi \widetilde{u} \, dx \qquad \text{for all } \varphi \in H^1_0(\Omega), \tag{123}$$

where $\chi = (w_1, w_2)$. Let us assume, that $\chi \in H(\mathrm{div}, \Omega)$. Then due to

$$-\int_\Omega \mathit{div}\ \chi \cdot \varphi \, dx = \int_\Omega \nabla\varphi \cdot \chi \, dx,$$

(123) is equivalent to

$$\int_\Omega (-\mathit{div}\ \chi + \gamma w_3) \cdot \varphi \, dx = \int_\Omega \widetilde{u}\varphi \, dx \qquad \text{for all } \varphi \in H^1_0(\Omega). \tag{124}$$

Equation (124) is equivalent to

$$-\mathit{div}\ \chi + \gamma w_3 = \widetilde{u}.$$

Therefore we can choose

$$w_3 = \frac{1}{\gamma}(\mathit{div}\ \chi + \widetilde{u}), \tag{125}$$

and determine χ in $H(\mathrm{div}, \Omega)$ "free". Due to the remarks after Theorem 6.9 and Remark 6.11 the optimal choice would be $w = \frac{1}{\nu_e} T u_e = \frac{1}{\nu_e}(\nabla u_e, u_e)$. Thus we choose $\chi \in (H^1(\Omega))^2$ satisfying

$$\chi \approx \frac{1}{\nu} \nabla \widetilde{u}.$$

Then $\chi \in H(\mathrm{div}, \Omega)$ is indeed satisfied. (Recall, that we compute χ via continuous finite elements, i.e., $\chi \in (H^1(\Omega))^2$ is "automatically" satisfied.)

At this point it became clear, why we need the "artificial" constant γ: without γ we would not have the freedom to choose χ. We would have to solve then

$$\int_\Omega -div\,\chi \cdot \varphi\, dx = \int_\Omega \widetilde{u}\varphi\, dx \qquad \text{for all } \varphi \in H_0^1(\Omega)$$

exactly, which is a hard task.

As we are looking for the smallest eigenvalue ν_1, we are left to find $\beta \in \mathbb{R}$ such that $\lambda_1 < \beta \leq \lambda_2$ (or $\lambda_n < \beta \leq \lambda_{n+1}$ for some $n \geq 1$) holds. To fulfil this task, we can use a domain homotopy. As the exact eigenvalues of the Laplace operator are known on rectangular domains, we can choose for Ω_0 a rectangle such that $\Omega \subset \Omega_0$.

Examples:

1. *Disc-like domain $\Omega_\circ$:* For the description of this domain see Section 4.5, Example 2 and Figure 5. In this case we need only one homotopy step. Let us choose $\gamma = 1$. The base domain Ω_0 is a square with side length $\frac{11}{8}$, that contains $\Omega_\circ$. The second Dirichlet-eigenvalue of the Laplace operator on Ω_0 is $\mu_2 = \frac{320\pi^2}{121} \geq 26$, thus $\underline{\nu}_2 = 26 + \gamma = 27$. The Rayleigh-Ritz upper bound for the first eigenvalue of (122) on $\Omega_\circ$ is $\overline{\nu}_1 \leq 12.9277$. Thus we can choose $\beta = 27$. Now Corollary 6.10 yields the lower bound $\nu_1 \geq \underline{\nu}_1 = 12.9122$, and thus $\mu_1 \geq \underline{\nu}_1 - \gamma = 11.9122$.

2. *Dumbbell-like domain Ω_d:* For the description of this domain see Section 4.5, Example 3 and Figure 9. In this case we need only one step as well. Let the base domain Ω_0 be the rectangle with corners $(0, -2.1)$, $(0, 2.1)$, $(10, 2.1)$, $(10, -2.1)$ and let $\gamma = 1$. The first two eigenvalues of (122) are not well separated, as we can see it on the Rayleigh-Ritz upper bounds $\overline{\nu}_1 = 2.38192$ and $\overline{\nu}_2 = 2.39065$. Therefore we have to use Theorem 6.9 with $n = 2$, instead of Corollary 6.10. The third smallest eigenvalue of the Laplace operator on

Ω_0 is known, thus

$$\nu_3 \geq \nu_3^{(0)} = \gamma + \pi^2 \left(\frac{1}{4.2^2} + \frac{9}{100} \right) \geq 2.445 > \overline{\nu}_2 = 2.39065 \geq \nu_2.$$

Hence we can apply Theorem 6.9 with $\beta = 2.445$ and $n = 2$. This yields us lower bounds $\underline{\nu}_1 = 2.31832$ and $\underline{\nu}_2 = 2.34139$. Therefore we obtain the enclosing intervals $[1.31832, 1.38192]$ for the first and $[1.34139, 1.39065]$ for the second eigenvalue of the Laplace operator on Ω_d.

6.3.5 Application to problem $\Delta^2 u + \alpha u = \nu(\widetilde{c}(x) + \alpha)u$

In this section we are aiming at bounds for the eigenvalues of problem (67) near to 1, i.e., of

$$\int_\Omega \Delta u \Delta \varphi \, dx + \alpha \int_\Omega u\varphi \, dx = \nu \int_\Omega (\widetilde{c}(x) + \alpha) u\varphi \, dx \quad \text{for all } \varphi \in H_0^2(\Omega), \quad (126)$$

with $u \in H_0^2(\Omega)$, $\alpha > -\min_{x\in\Omega} \widetilde{c}(x)$ and $\alpha > 0$. Observe that $\kappa = \frac{1}{\nu}$ and $\mu = 1 - \kappa = 1 - \frac{1}{\nu}$ with κ from (67) and μ from (66).

We apply the results of Section 6.3.1 with the following casting: the Hilbert space $H = H_0^2(\Omega)$ provided with the inner product

$$\langle \varphi, \psi \rangle_{H_0^2,\alpha} = \int_\Omega \Delta\varphi \Delta\psi \, dx + \alpha \int_\Omega \varphi\psi \, dx.$$

The right-hand side $N(\varphi, \psi) = \int_\Omega (\widetilde{c}(x) + \alpha)\varphi\psi \, dx$.

Let us define the Hilbert space $(X, b(\cdot,\cdot))$ and the isometry $T : H \to X$ as follows:

$$X = \left(L_2(\Omega)\right)^2, \qquad T\varphi = (\Delta\varphi, \varphi) \quad \text{for all } \varphi \in H_0^2(\Omega),$$

$$b(\varphi, \psi) = \langle \varphi_1, \psi_1 \rangle_{L_2} + \alpha \langle \varphi_2, \psi_2 \rangle_{L_2} \quad \text{for all } \varphi = (\varphi_1, \varphi_2), \psi = (\psi_1, \psi_2) \in X.$$

Then the assumption $b(T\varphi, T\psi) = \langle \varphi, \psi \rangle_{H^2_{0,\alpha}}$ is obviously satisfied.

Now let $(\hat{u}_1, \nu_1), \ldots, (\hat{u}_n, \nu_n) \in H_0^2(\Omega) \times \mathbb{R}$ be approximate eigenpairs to the exact eigenpairs
$(u_1^e, \nu_1^e), \ldots, (u_n^e, \nu_n^e) \in H_0^2(\Omega) \times \mathbb{R}$. We can choose the vectors $w_i = (w_i^1, w_i^2) \in X$, $(i = 1, \ldots, n)$ satisfying (103) in the following way: Condition (103) reads in this context

$$\int_\Omega w_i^1 \cdot \Delta\varphi \, dx + \alpha \int_\Omega w_i^2 \varphi \, dx = \int_\Omega (\tilde{c}(x) + \alpha) \hat{u}_i \varphi \, dx \quad \text{for } \varphi \in H_0^2(\Omega), 1 \le i \le n. \tag{127}$$

If $\Delta w_i^1 \in L_2(\Omega)$, then (127) is equivalent to

$$\int_\Omega \Delta w_i^1 \cdot \varphi \, dx + \alpha \int_\Omega w_i^2 \varphi \, dx = \int_\Omega (\tilde{c}(x) + \alpha) \hat{u}_i \varphi \, dx \quad \text{for } \varphi \in H_0^2(\Omega), 1 \le i \le n, \tag{128}$$

which is equivalent to

$$\Delta w_i^1 + \alpha w_i^2 = (\tilde{c}(x) + \alpha) \hat{u}_i, \tag{129}$$

for $i = 1, \ldots, n$. Therefore we define

$$w_i^2 = \frac{1}{\alpha}((\tilde{c}(x) + \alpha) \hat{u}_i - \Delta w_i^1), \tag{130}$$

and then we are free to determine w_i^1 with $\Delta w_i^1 \in L_2(\Omega)$, $(i = 1, \ldots, n)$. Similarly to the role of γ in Section 6.3.4 the role of α is to have the freedom to choose w_i^1. Otherwise we would have to solve

$$\int_\Omega \Delta w_i^1 \cdot \varphi \, dx = \int_\Omega \tilde{c}(x) \hat{u}_i \varphi \, dx \qquad \text{for all } \varphi \in H_0^2(\Omega), \quad i = 1, \ldots, n,$$

exactly, which is a hard task.

Corresponding to the optimal choice $w_i = \frac{1}{\nu_i^e} T u_i^e = \frac{1}{\nu_i^e} (\Delta u_i^e, u_i^e)$, we need to choose

$$w_i^1 \in L_2(\Omega) \quad \text{such that} \quad w_i^1 \approx \frac{1}{\nu_i}\Delta\hat{u}_i \quad \text{and} \quad \Delta w_i^1 \in L_2(\Omega), \quad i = 1, \ldots, n. \tag{131}$$

The problem of computing w_i is connected to the computation of the approximate eigenfunction $\hat{u}_i$. Let us discuss this second problem first, and then we come back to the definition of w_i.

By the computation of the approximate eigenfunctions $\hat{u}_i \in H_0^2(\Omega)$ we have the same difficulty, as we had for the approximate solution $\omega \in H_0^2(\Omega)$ of (36): since we use only continuous, i.e., H^1-finite element approximations, in general we can not compute H^2-functions. We can solve this problem again with an analogous technique as we used earlier for ω in Section 3.2.2. We compute numerical approximations to u_i^e, ∇u_i^e, $-\Delta u_i^e$ and $-\nabla\Delta u_i^e$, i.e.,

$$\begin{aligned}
\widetilde{u}_i &\approx u_i^e, & \widetilde{u}_i &\in H_0^1(\Omega)\\
\widetilde{\sigma}_i &\approx \nabla u_i^e, & \widetilde{\sigma}_i &\in \left(H_0^1(\Omega)\right)^2\\
\widetilde{v}_i &\approx -\Delta u_i^e, & \widetilde{v}_i &\in H^1(\Omega)\\
\widetilde{\rho}_i &\approx -\nabla\Delta u_i^e, & \widetilde{\rho}_i &\in H(\text{div}, \Omega),
\end{aligned}$$

and approximate eigenvalues $\widetilde{\nu}_i$ to ν_i^e for $i = 1, \ldots, n$. For these computations continuous finite elements are sufficient. Let $\hat{u}_i \in H_0^2(\Omega)$ be defined (not actually computed) via

$$\Delta^2\hat{u}_i = \Delta div\, \widetilde{\sigma}_i \quad \text{on } \Omega,$$

i.e.,

$$\langle \Delta\hat{u}_i, \Delta\varphi\rangle_{L_2} = \langle div\, \widetilde{\sigma}_i, \Delta\varphi\rangle_{L_2} \qquad \text{for all } \varphi \in H_0^2(\Omega),$$

for $i = 1, \ldots, n$.

Furthermore, let us introduce the following notation:

$$\hat{\sigma}_i = \nabla \hat{u}_i \in (H_0^1(\Omega))^2, \qquad i = 1, \ldots, n. \tag{132}$$

Observe, that again because of the definition

$$rot\ \hat{\sigma}_i = 0, \qquad i = 1, \ldots, n. \tag{133}$$

With the help of the above auxiliary functions we can also define w_1^i that meet the requirements:

$$\begin{aligned} -\Delta w_i^1 &= \frac{1}{\widetilde{\nu}_i} div\, \widetilde{\rho}_i \quad \text{on } \Omega, \\ w_i^1 &= -\frac{1}{\widetilde{\nu}_i} \widetilde{v}_i \quad \text{on } \partial\Omega, \end{aligned}$$

i.e.,

$$\int_\Omega \nabla z \nabla \varphi\, dx = \frac{1}{\widetilde{\nu}_i} \int_\Omega \text{div}\, \widetilde{\rho}_i \varphi + \nabla \widetilde{v}_i \nabla \varphi\, dx \qquad \text{for all } \varphi \in H_0^1(\Omega),$$

with $z \in H_0^1(\Omega)$ and $w_i^1 = z - \frac{1}{\widetilde{\nu}_i}\widetilde{v}_i$, for $i = 1, \ldots, n$. From this definition follows, that $w_i^1 \in H^1(\Omega)$ and $\Delta w_i^1 \in L_2(\Omega)$, but in general $w_1^i \notin H^2(\Omega)$, $(i = 1, \ldots, n)$.

For the method of Lehmann-Goerisch as well as for the method of Rayleigh-Ritz the matrices

$$A_0 = \Big(\int_\Omega \Delta \hat{u}_i \Delta \hat{u}_j\, dx + \alpha \int_\Omega \hat{u}_i \hat{u}_j\, dx \Big)_{i,j=1,\ldots,n},$$

$$A_1 = \Big(\int_\Omega (\alpha + \widetilde{c}(x)) \hat{u}_i \hat{u}_j\, dx \Big)_{i,j=1,\ldots,n},$$

$$A_2 = \Big(\int_\Omega w_i^1 w_j^1\, dx + \alpha \int_\Omega w_i^2 w_j^2\, dx \Big)_{i,j=1,\ldots,n}$$

are needed. The entries a_{ij}^k, $(i,j = 1,\dots,n,\ k = 0,1,2)$ of these matrices contain the functions w_i and $\hat{u}_i$ $(i = 1,\dots,n)$, which are not computed. Therefore, using the approximations

$$
\begin{aligned}
\hat{u}_i &\approx \widetilde{u}_i, \\
\Delta\hat{u}_i &\approx \mathrm{div}\ \widetilde{\sigma}_i, \\
w_i^1 &\approx -\frac{1}{\widetilde{\nu}_i}\widetilde{v}_i, \\
w_i^2 &\approx \frac{1}{\alpha}\left((\widetilde{c}(x)+\alpha)\widetilde{u}_i + \frac{1}{\widetilde{\nu}_i}\mathrm{div}\ \widetilde{\rho}_i\right)
\end{aligned}
$$

instead of A_0, A_1, A_2 we compute the following approximate matrices

$$\widetilde{A}_0 = \Big(\int_\Omega \mathrm{div}\ \widetilde{\sigma}_i\ \mathrm{div}\ \widetilde{\sigma}_j\ dx + \alpha\int_\Omega \widetilde{u}_i\widetilde{u}_j\ dx\Big)_{i,j=1,\dots,n},$$

$$\widetilde{A}_1 = \Big(\int_\Omega (\alpha+\widetilde{c}(x))\widetilde{u}_i\widetilde{u}_j\ dx\Big)_{i,j=1,\dots,n},$$

$$\widetilde{A}_2 = \Big(\frac{1}{\widetilde{\nu}_i\widetilde{\nu}_j}\int_\Omega \widetilde{v}_i\widetilde{v}_j\ dx +$$

$$+\frac{1}{\alpha}\int_\Omega((\alpha+\widetilde{c}(x))\widetilde{u}_i + \frac{1}{\widetilde{\nu}_i}\mathrm{div}\ \widetilde{\rho}_i)((\alpha+\widetilde{c}(x))\widetilde{u}_j + \frac{1}{\widetilde{\nu}_j}\mathrm{div}\ \widetilde{\rho}_j)\ dx\Big)_{i,j=1,\dots,n},$$

with entries $\widetilde{a}_{ij}^k$, $(i,j = 1,\dots,n,\ k = 0,1,2)$. By calculating upper estimates for the errors

$$|a_{ij}^k - \widetilde{a}_{ij}^k| \le \varepsilon_{ij}^k, \qquad i,j = 1,\dots,n,\ k = 0,1,2,$$

we can enclose the matrices A_0, A_1, A_2 into the interval matrices

$$A_0 \in [\widetilde{A}_0^{ij} - \varepsilon_0^{ij}, \widetilde{A}_0^{ij} + \varepsilon_0^{ij}] =: \overline{A_0},$$

$$A_1 \in [\widetilde{A}_1^{ij} - \varepsilon_1^{ij}, \widetilde{A}_1^{ij} + \varepsilon_1^{ij}] =: \overline{A_1},$$

$$A_2 \in [\widetilde{A}_2^{ij} - \varepsilon_2^{ij}, \widetilde{A}_2^{ij} + \varepsilon_2^{ij}] =: \overline{A_2}.$$

For this purpose we calculate estimates analogous to (43), (44) and (45), i.e.,

$$\|\Delta \hat{u}_i - div\ \widetilde{\sigma}_i\|_{L_2} = \|div\ \hat{\sigma}_i - div\ \widetilde{\sigma}_i\|_{L_2} \leq D \cdot \|rot\ \widetilde{\sigma}_i\|_{L_2}, \tag{134}$$

$$\|\nabla \hat{u}_i - \widetilde{\sigma}_i\|_{L_2} \leq C_{H_0^1 \hookrightarrow L_2} \cdot \sqrt{D^2+1} \cdot \|rot\ \widetilde{\sigma}_i\|_{L_2}, \tag{135}$$

$$\|\nabla \hat{u}_i - \nabla \widetilde{u}_i\|_{L_2} \leq \|\widetilde{\sigma}_i - \nabla \widetilde{u}_i\|_{L_2} + C_{H_0^1 \hookrightarrow L_2} \cdot \sqrt{D^2+1} \cdot \|rot\ \widetilde{\sigma}_i\|_{L_2}, \tag{136}$$

$$\|\hat{u}_i - \widetilde{u}_i\|_{L_2} \leq C_{H_0^1 \hookrightarrow L_2} \cdot \left(C_{H_0^1 \hookrightarrow L_2} \cdot \sqrt{D^2+1} \cdot \|rot\ \widetilde{\sigma}_i\|_{L_2} + \|\widetilde{\sigma}_i - \nabla \widetilde{u}_i\|_{L_2} \right), \tag{137}$$

for $i = 1, \ldots, n$. Moreover, using that for $g \in H_0^1(\Omega)$

$$\|\Delta g\|_{H^{-1}} = \sup_{\varphi \in H_0^1(\Omega), \varphi \neq 0} \frac{|\int_\Omega \nabla g \nabla \varphi\, dx|}{\|\nabla \varphi\|_{L_2}} = \|\nabla g\|_{L_2}$$

holds, we have for $i = 1, \ldots, n$ that

$$\|w_i^1 + \frac{1}{\widetilde{\nu}_i} \widetilde{v}_i\|_{L_2} \leq C_{H_0^1 \hookrightarrow L_2} \|\nabla w_i^1 + \frac{1}{\widetilde{\nu}_i} \nabla \widetilde{v}_i\|_{L_2} = C_{H_0^1 \hookrightarrow L_2} \|\Delta w_i^1 + \frac{1}{\widetilde{\nu}_i} \Delta \widetilde{v}_i\|_{H^{-1}}$$

$$= \frac{1}{\widetilde{\nu}_i} C_{H_0^1 \hookrightarrow L_2} \|div\ \widetilde{\rho}_i - \Delta \widetilde{v}_i\|_{H^{-1}} \overset{(46)}{\leq} \frac{1}{\widetilde{\nu}_i} C_{H_0^1 \hookrightarrow L_2} \|\widetilde{\rho}_i - \nabla \widetilde{v}_i\|_{L_2}. \tag{138}$$

At last we calculate

$$\left\| w_i^2 - \frac{1}{\alpha} \left((\widetilde{c}(x) + \alpha) \widetilde{u}_i + \frac{1}{\widetilde{\nu}_i} \text{div}\ \widetilde{\rho}_i \right) \right\|_{L_2} = \frac{1}{\alpha} \|(\widetilde{c}(x) + \alpha)(\widetilde{u}_i - \hat{u}_i)\|_{L_2}$$

$$\leq \frac{1}{\alpha} (\alpha + \max_{x \in \Omega} \widetilde{c}(x)) \|\widetilde{u}_i - \hat{u}_i\|_{L_2}, \qquad i = 1, \ldots, n. \tag{139}$$

Observe that all the terms on the right-hand sides of the above inequalities are computable quantities, if we use (137) in (139).

With the help of these estimates we are able now to compute ε_{ij}^k by using equality

$$ab - cd = (a-c)(b-d) + c(b-d) + d(a-c)$$

and the estimates (134) to (139). For example:

$$|a_{ij}^0 - \widetilde{a}_{ij}^0| = \left| \int_\Omega \Delta\hat{u}_i \Delta\hat{u}_j - div\ \widetilde{\sigma}_i\ div\ \widetilde{\sigma}_j\ dx + \alpha \int_\Omega \hat{u}_i\hat{u}_j - \widetilde{u}_i\widetilde{u}_j\ dx \right|$$

$$\leq \|\Delta\hat{u}_i - div\ \widetilde{\sigma}_i\|_{L_2} \|\Delta\hat{u}_j - div\ \widetilde{\sigma}_j\|_{L_2} +$$

$$+ \|div\ \widetilde{\sigma}_i\|_{L_2} \|\Delta\hat{u}_j - div\ \widetilde{\sigma}_j\|_{L_2} + \|div\ \widetilde{\sigma}_j\|_{L_2} \|\Delta\hat{u}_i - div\ \widetilde{\sigma}_i\|_{L_2}$$

$$+ \alpha \left(\|\hat{u}_i - \widetilde{u}_i\|_{L_2} \|\hat{u}_j - \widetilde{u}_j\|_{L_2} + \|\widetilde{u}_i\|_{L_2} \|\hat{u}_j - \widetilde{u}_j\|_{L_2} + \|\widetilde{u}_j\|_{L_2} \|\hat{u}_i - \widetilde{u}_i\|_{L_2} \right)$$

$$\leq D^2 \|rot\ \widetilde{\sigma}_i\|_{L_2} \|rot\ \widetilde{\sigma}_j\|_{L_2} + D \|div\ \widetilde{\sigma}_i\|_{L_2} \|rot\ \widetilde{\sigma}_j\|_{L_2} + D \|div\ \widetilde{\sigma}_j\|_{L_2} \|rot\ \widetilde{\sigma}_i\|_{L_2}$$

$$+ \alpha \Big[C^2_{H_0^1 \hookrightarrow L_2} \left(C_{H_0^1 \hookrightarrow L_2} \sqrt{D^2+1} \|rot\ \widetilde{\sigma}_i\|_{L_2} + \|\widetilde{\sigma}_i - \nabla\widetilde{u}_i\|_{L_2} \right)$$

$$\cdot \left(C_{H_0^1 \hookrightarrow L_2} \sqrt{D^2+1} \|rot\ \widetilde{\sigma}_j\|_{L_2} + \|\widetilde{\sigma}_j - \nabla\widetilde{u}_j\|_{L_2} \right)$$

$$+ \|\widetilde{u}_i\|_{L_2} C_{H_0^1 \hookrightarrow L_2} \cdot \left(C_{H_0^1 \hookrightarrow L_2} \cdot \sqrt{D^2+1} \cdot \|rot\ \widetilde{\sigma}_j\|_{L_2} + \|\widetilde{\sigma}_j - \nabla\widetilde{u}_j\|_{L_2} \right)$$

$$+ \|\widetilde{u}_j\|_{L_2} C_{H_0^1 \hookrightarrow L_2} \cdot \left(C_{H_0^1 \hookrightarrow L_2} \cdot \sqrt{D^2+1} \cdot \|rot\ \widetilde{\sigma}_i\|_{L_2} + \|\widetilde{\sigma}_i - \nabla\widetilde{u}_i\|_{L_2} \right) \Big].$$

For the Rayleigh-Ritz method one needs the eigenvalues of the matrix eigenvalue problem

$$A_0 x = \lambda A_1 x, \tag{140}$$

while for the Lehmann-Goerisch method the eigenvalues of

$$(A_0 - \beta A_1)x = \widetilde{\lambda}(A_0 - 2\beta A_1 + \beta^2 A_2)x. \tag{141}$$

We can obtain an enclosure for the above eigenvalues by means of the interval matrix eigenvalue problems

$$\overline{A}_0 x = \Lambda \overline{A}_1 x, \tag{142}$$

$$(\overline{A}_0 - \beta\overline{A}_1)x = \widetilde{\Lambda}(\overline{A}_0 - 2\beta\overline{A}_1 + \beta^2\overline{A}_2)x. \tag{143}$$

If the matrices are of small dimension ($n = 1, 2$), then we can obtain such an enclosure rather directly. For higher dimensions we can use the following Lemma of Plum, see [13].

6.12 Lemma *Let $\overline{A}$, $\overline{B} \in \mathbb{C}^{n,n}$ be complex Hermitian interval matrices such that B is positive definite for all $B \in \overline{B}$. For some fixed Hermitian $A_0 \in \overline{A}$ and $B_0 \in \overline{B}$, let $(\widetilde{\lambda}_k, \widetilde{x}_k)$, $(k = 1, \ldots, n)$ denote approximate eigenpairs of $A_0x = \lambda B_0x$ with $\widetilde{x}_k^T B_0 \widetilde{x}_l \approx \delta_{k,l}$.*

Suppose that for some r_0, $r_1 > 0$,

$$\|X^TAX - X^TBX\Lambda\|_\infty \le r_0 \quad \|X^TBX - I\|_\infty \le r_1 \quad \text{for all } A \in \overline{A},\ B \in \overline{B},$$

where $X = (\widetilde{x}_1, \ldots, \widetilde{x}_n)$, $\Lambda = \operatorname{diag}(\widetilde{\lambda}_1, \ldots, \lambda_n)$. If $r_1 < 1$, we have for all $A \in \overline{A}$, $B \in \overline{B}$ and all eigenvalues λ of $Ax = \lambda Bx$

$$\lambda \in \bigcup_{k=1}^{n} B_r(\widetilde{\lambda}_k) \quad \text{with } r = \frac{r_0}{1 - r_1}, \qquad i = 1, \ldots, n,$$

where $B_r(\lambda) = \{z \in \mathbb{C} \colon |z - \lambda| \le r\}$. Moreover, each connected component of this union contains as many eigenvalues as midpoints $\widetilde{\lambda}_k$. Especially, if the balls $B_r(\widetilde{\lambda}_k)$ are disjoint, then we have $\lambda_k \in B_r(\widetilde{\lambda}_k)$ for $k = 1, \ldots, n$.

In case of the Rayleigh-Ritz method we obtain enclosures $\lambda_i \in \mathcal{L}_i$ ($i = 1, \ldots, n$) for the n eigenvalues $\lambda_1, \ldots, \lambda_n$ of (140). Theorem 6.8 yields us then the desired upper bounds for the n smallest eigenvalues $\kappa_1, \ldots, \kappa_n$ of (126)

$$\kappa_i \le \lambda_i \le \sup \mathcal{L}_i, \qquad i = 1, \ldots, n.$$

In case of the Lehmann-Goerisch method we obtain the enclosures $\widetilde{\lambda}_i \in \widetilde{\mathcal{L}}_i$ ($i = 1, \ldots, n$) for the n eigenvalues $\widetilde{\lambda}_1, \ldots, \widetilde{\lambda}_n$ of (141). Theorem 6.9 yields us then the desired lower bounds for the n largest eigenvalues $\kappa_{m-n} \le \cdots \le \kappa_m$

of (126) below β,

$$\kappa_i \geq \beta - \frac{\beta}{1-\widetilde{\lambda}_i} \geq \beta - \frac{\beta}{1-\sup \widetilde{\mathcal{L}}_i}, \qquad i = m-n, \ldots, m.$$

Homotopy for finding β

At last in this section we deal with the problem of finding β, which satisfies assumptions (i) and (ii) of Theorem 6.9 for problem (126).

Problem (126) is similar to the eigenvalue problem for the biharmonic operator in the sense that on the left-hand-side of (126) stands also the biharmonic operator. Enclosures for the first 100 Dirichlet eigenvalues of the biharmonic operator are known on the unit square $E_\square$, i.e. for

$$\int_{E_\square} \Delta u \Delta \varphi \, dx = \nu \int_{E_\square} u \varphi \, dx, \qquad \text{for all } \varphi \in H_0^2(E_\square),$$

see [31]. Thus we can start our homotopy with a domain homotopy connecting a square Ω_0 containing the domain $\Omega = \Omega_1$ with Ω_1. We gave a more detailed description of the domain homotopy in Section 6.3.3. At the end of this homotopy we gain lower bounds for the k smallest eigenvalues of the biharmonic operator on Ω, for some suitable k.

Afterwards we start a linear homotopy on the domain Ω connecting the eigenvalue problem for the biharmonic operator with problem (126) via intermediate eigenvalue problems. More precisely let $M = \max_{x \in \overline{\Omega}} \widetilde{c}(x)$ and let the base problem be

$$\int_\Omega \Delta u \Delta \varphi + \alpha u \varphi \, dx = \lambda^{(0)}(M+\alpha) \int_\Omega u \varphi \, dx, \qquad \text{for all } \varphi \in H_0^2(\Omega), \quad (144)$$

the eigenvalues of which can clearly be computed from the Dirichlet eigenvalues of the biharmonic operator. This means with the notations of Section

6.3.3 that $H = H_0^2(\Omega)$,

$$N_0(u,\varphi) = (M+\alpha)\int_\Omega u\varphi\,dx, \quad \text{and} \quad N_1(u,\varphi) = \int_\Omega (\widetilde{c}(x)+\alpha)u\varphi\,dx.$$

Assumption (120) is obviously satisfied. Thus the intermediate eigenvalue problems for $s \in [0,1]$ are

$$\int_\Omega \Delta u \Delta\varphi + \alpha u\varphi\,dx \tag{145}$$
$$= \lambda^{(s)} \int_\Omega \left((1-s)(M+\alpha) + s(\widetilde{c}(x)+\alpha)\right) u\varphi\,dx, \quad \text{for all } \varphi \in H_0^2(\Omega).$$

As explained in the general description in Section 6.3.3, the lower bounds obtained in the last step in the linear homotopy is an appropriate value for β we were looking for.

6.4 Enclosure of integrals

In this section we give bounds for integral-expressions appearing in our work.

6.4.1 Enclosure for the moments M_s

In Section 3.4 we defined the moments of a domain Q as

$$M_s = \max_{x_0\in Q}\left[\frac{1}{|Q|}\int_Q |x-x_0|^s\,dx\right]^{\frac{1}{s}}.$$

In the case, when Q is a circular disc or a square, and p is even, one can calculate M_s in closed form. But for general s we need a cubature formula with remainder term, or we can treat this problem as follows. If Q is a circular disc with midpoint 0 and radius r, then

$$M_s(Q) = \left[\frac{1}{|Q|}\int_Q |x-(r,0)|^s\,dx\right]^{\frac{1}{s}}.$$

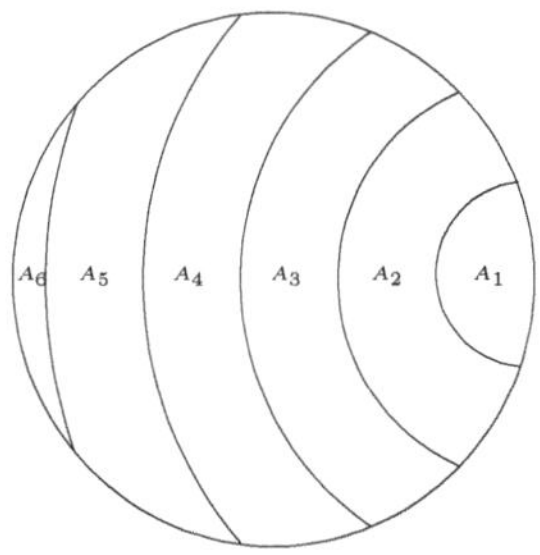

Figure 29: *Division of a circular disc into circle-strips*

The function $x \to |x-(r,0)|^s$ is constant along circles with center $(r,0)$, and monotone increasing as the radius of these circles grows. Thus we can split Q in circle-strips A_i, as shown in Figure 29 and

$$\int_Q |x-(r,0)|^s \, dx \leq \sum_{i=1}^{n} T_i q_i^s,$$

where q_i denotes the major radius of the circle-strip A_i, and T_i denotes its area. If n is big enough, we get a good upper bound for the integral.

In case of a square as shown on Figure 30 we can proceed analogously. Now we have

$$M_s(Q) = \left[\frac{1}{|Q|}\int_Q |x|^s \, dx\right]^{\frac{1}{s}}.$$

We can divide the square into circle-strips as shown in Figure 30.

The advantage of this method is that we can omit the supremum norm of some derivatives of the integrand, as usually required for validated cubature formulas.

6.4.2 A cubature formula with computable error term

We use isoparametric elements to represent the domain Ω. This fact has amongst others the consequence, that the transformation of the unit triangle

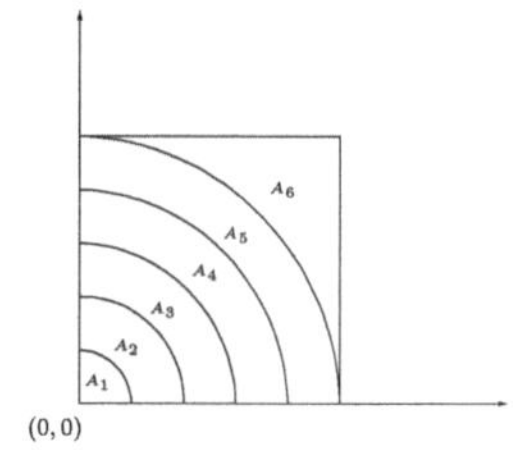

Figure 30: *Division of a square into circle-strips*

onto each of the elements has the form

$$T(\xi,\eta) = (P_1(\xi,\eta), P_2(\xi,\eta)),$$

where P_1, P_2 are cubic polynomials. If we calculate integrals of derivatives of some finite element function, then we have to deal with a rational expression. In general such expressions can not be integrated in closed form. Therefore we need a cubature formula with computable error term. We use for the error representation the so-called Sard kernels. We give a short description of the theory, for more detail see [8].

Let Ω_0 denote the unit triangle throughout this chapter. Let $C^k_{\Omega_0}$ be a cubature formula on Ω_0, which is exact for all polynomials of degree k. Let $E^{x,y}_{\Omega_0}$ denote the error of the cubature, i.e., for a function $f\colon \Omega_0 \to \mathbb{R}$ let

$$E^{x,y}_{\Omega_0}(f) = \int_{\Omega_0} f(x,y)\, d(x,y) - C_{\Omega_0}(f).$$

Further, we use the notation

$$y_+ = \begin{cases} y & \text{if } 0 \le y \\ 0 & \text{otherwise} \end{cases},$$

and for $i, j \ge 0$

$$f_{i,j}(x,y) = \frac{\partial^{i+j} f(x,y)}{\partial x^i \partial y^j}.$$

6.13 Definition *Let $k > 0$ and $k = 2l$ or $k = 2l + 1$. Define for $(x, y) \in \overline{\Omega}$*

$$\begin{aligned}
K_1^i(u) &= E_{\Omega_0}^{x,y}[y^i(x-u)_+^{k-i}] \qquad \text{for } i = 0, \ldots, l,\\
K_2^i(v) &= E_{\Omega_0}^{x,y}[x^i(y-v)_+^{k-i}] \qquad \text{for } i = 0, \ldots, l,\\
K_{l,l}(u,v) &= E_{\Omega_0}^{x,y}[(x-u)_+^l(y-v)_+^l] \qquad \text{if } k = 2l,\\
K_{l,l-1}(u,v) &= E_{\Omega_0}^{x,y}[(x-u)_+^l(y-v)_+^{l-1}] \qquad \text{if } k = 2l+1,\\
K_{l-1,l}(u,v) &= E_{\Omega_0}^{x,y}[(x-u)_+^{l-1}(y-v)_+^l] \qquad \text{if } k = 2l+1.
\end{aligned}$$

These functions are called the Sard kernel functions of $E_{\Omega_0}^{x,y}$.

6.14 Theorem *Let $k > 0$. For $f \in C^{k+1}(\overline{\Omega}_0)$ we have*

$$E_{\Omega_0}^{x,y}(f) = \frac{1}{k!}\sum_{i=0}^{l}\binom{k}{i}\Big[\int_0^1 K_1^i(u) f_{k+1-i,i}(u,0)\,du + \int_0^1 K_2^i(v) f_{i,k+1-i}(0,v)\,dv\Big]$$

$$+\frac{1}{l!l!}\int_{\Omega_0} K_{l,l}(u,v) f_{l+1,l+1}(u,v)\,d(u,v),$$

if $k = 2l + 1$ and

$$E_{\Omega_0}^{x,y}(f) = \frac{1}{2k!}\Big[\int_0^1 K_1^0(u) f_{k+1,0}(u,0)\,du + \int_0^1 K_2^0(v) f_{0,k+1}(0,v)\,dv\Big]$$

$$+\frac{1}{k!}\sum_{i=0}^{l}\binom{k}{i}\Big[\int_0^1 K_1^i(u) f_{k+1-i,i}(u,0)\,du + \int_0^1 K_2^i(v) f_{i,k+1-i}(0,v)\,dv\Big]$$

$$+\frac{1}{2(l-1)!l!}\Big[\int_{\Omega_0} K_{l,l-1}(u,v) f_{l+1,l}(u,v)\,d(u,v) +$$

$$+\int_{\Omega_0} K_{l-1,l}(u,v) f_{l,l+1}(u,v)\,d(u,v)\Big],$$

if $k = 2l$.

With the help of Theorem 6.14 we obtain an upper bound for $|E_{\Omega_0}^{x,y}(f)|$, if we compute upper bounds for the supremum norm of the derivatives of f and for the integrals of the absolute value of the Sard kernels occurring in the theorem.

We use the following cubature formula on Ω_0. Let $P_0 = (\frac{1}{2}, 0)$, $P_1 = (\frac{1}{2}, \frac{1}{2})$, $P_2 = (0, \frac{1}{2})$. Let the cubature C_{Ω_0} be defined for a function $f\colon \Omega_0 \to \mathbb{R}$ via

$$C_{\Omega_0}(f) = \frac{1}{6}(f(P_0) + f(P_1) + f(P_2))$$

This formula is exact for all polynomials of degree 2, i.e., $k = 2$, $l = 1$ in the theorem. If we calculate the integrals of the Sard kernels listed above, then we get the desired upper bound:

$$|E_{\Omega_0}^{x,y}(f)| \leq 0.0013889 \cdot (\|f_{3,0}(u,0)\|_{\infty,[0,1]} + \|f_{0,3}(0,v)\|_{\infty,[0,1]})$$

$$+0.0014215 \cdot (\|f_{2,1}(u,0)\|_{\infty,[0,1]} + \|f_{1,2}(0,v)\|_{\infty,[0,1]})$$

$$+0.00290425 \cdot (\|f_{2,1}(u,v)\|_{\infty,\Omega_0} + \|f_{1,2}(u,v)\|_{\infty,\Omega_0}).$$

To increase the precision of the integrals, we divide first the unit triangle into four subdomains as it can be seen in Figure 31, and we calculate the cubature and the error on each subdomains. If needed, we proceed iteratively with dividing each subtriangles as in the first step (Figure 31).

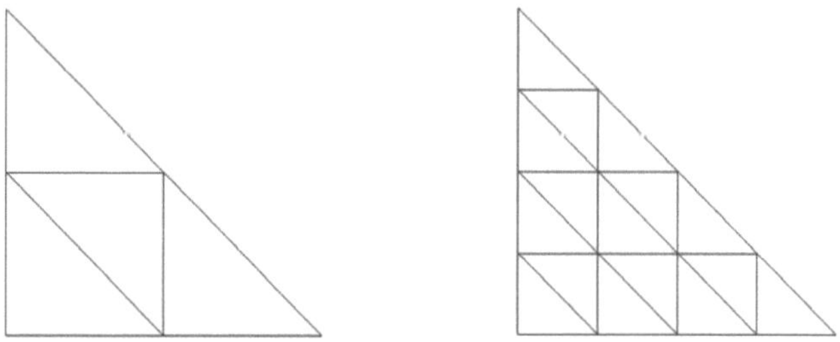

Figure 31: *Division of the unit triangle*

6.5 Numerical tools

All the computations were performed on the following computers:
1. InstitutsCluster of the Steinbuch Centre for Computing at KIT, with computing nodes with 2 Quad-Core Intel Xeon X5355 processor of 2,667 GHz and 16 GB central memory per dual board. For a more detailed description see the User's Guide of the Institutscluster [33].
2. High-performance Computer HP XC3000 of the Steinbuch Centre for Computing at KIT, with computing nodes with 2 Quad-Core Intel Xeon E5540 processor of 2,53 GHz and 24 GB central memory. For a more detailed description see the User's Guide of HPXC3000 [34].
3. Parallel computer Otto of the Institute for Applied and Numerical Mathematics of KIT with AMD Opteron 2352 processor of 2110.840 MHz and 32 GB central memory per dual board.
4. The cluster Taurus at the Engineering Mathematics and Computing Lab. of KIT with
5. Personal computers of the University of Debrecen and of KIT.

The computation of the approximate functions and the corresponding integrals were carried out with the C++-finite-elements-program M++ developed by C. Wieners and his research group. For a description of the package see [32]. We used continuous isoparametric cubic triangle-elements.

For solving linear systems we applied preconditioned GMRES with partial ILU preconditioner, and the direct solver SuperLU. For the approximate integrals we used a 7-degree quadrature rule. For computing eigenvalues of small matrices we used the linear algebra package Jama and the template numerical toolkit TNT.

For verified computing we calculated the appropriate expressions in interval arithmetics using the package C-XSC developed by R. Hammer, M. Hocks, U. Kulisch, D. Ratz et al.. For a detailed description see [12].

Notations

$\mathbb{R}_+$	–	the set of the positive real numbers
α	–	multiindex, $\alpha = (\alpha_1, \dots, \alpha_n) \in (\mathbb{N} \cup \{0\})^n$
$\lvert\alpha\rvert$	–	$\lvert\alpha\rvert = \alpha_1 + \cdots + \alpha_n$
Ω	–	domain in $\mathbb{R}^n$, (open and connected)
$\partial\Omega$	–	boundary of Ω
ν	–	exterior unit normal at $\partial\Omega$
$B_r(x)$	–	open ball with center in x and radius r
$\mathcal{B}(X,Y)$	–	the space of bounded linear operators from X to Y,
$C_0^\infty(\Omega)$	–	the space of smooth functions $f\colon \Omega \to \mathbb{R}$ with compact support in Ω
$L_p(\Omega)$	–	Lebesgue space
$\langle\cdot,\cdot\rangle_{L_2}$	–	inner product in $L_2(\Omega)$, $\langle u, v\rangle_{L_2} = \int_\Omega uv\, dx$
$\lVert\cdot\rVert_{L_p}$	–	norm in $L_p(\Omega)$
$\lVert\cdot\rVert_{L_2}$	–	norm in $(L_2(\Omega))^2$, $\lVert\sigma\rVert_{L_2}^2 = \lVert(\sigma_1,\sigma_2)\rVert_{L_2}^2 = \lVert\sigma_1\rVert_{L_2}^2 + \lVert\sigma_2\rVert_{L_2}^2$
$H^{m,p}(\Omega)$	–	Sobolev space of m-times weakly differentiable functions $f\colon \Omega \to \mathbb{R}$ with weak derivatives in $L_p(\Omega)$
$H^m(\Omega)$	–	$H^{m,2}(\Omega)$
$\lVert\cdot\rVert_{H^{m,p}}$	–	norm in $H^{m,p}(\Omega)$, $\lVert u\rVert_{H^{m,p}} = \Big(\sum_{\lvert\alpha\rvert\le m} \lVert D^\alpha u\rVert_{L_p}^p\Big)^{\frac{1}{p}}$
$H_0^{m,p}(\Omega)$	–	$\overline{C_0^\infty(\Omega)}$ in the norm of $H^{m,p}$ for bounded domains
$H_0^m(\Omega)$	–	$H_0^{m,2}(\Omega)$
$\langle\cdot,\cdot\rangle_{H_0^1}$	–	inner product in $H_0^1(\Omega)$, $\langle u, v\rangle_{H_0^1} = \int_\Omega \nabla u \nabla v\, dx$
$\lVert\cdot\rVert_{H_0^1}$	–	corresponding norm in $H_0^1(\Omega)$, $\lVert u\rVert_{H_0^1} = \lVert\nabla u\rVert_{L_2}$
$\langle\cdot,\cdot\rangle_{H_0^1}$	–	inner product in $\big(H_0^1(\Omega)\big)^2$, $\langle\sigma,\rho\rangle_{H_0^1} = \langle\sigma_1,\rho_1\rangle_{H_0^1} + \langle\sigma_2,\rho_2\rangle_{H_0^1}$

$\|\cdot\|_{H_0^1}$	–	corresponding norm in $\left(H_0^1(\Omega)\right)^2$,
		$\|\sigma\|_{H_0^1}^2 = \|(\sigma_1,\sigma_2)\|_{H_0^1}^2 = \|\sigma_1\|_{H_0^1}^2 + \|\sigma_2\|_{H_0^1}^2$
$\langle\cdot,\cdot\rangle_{H_0^1,\gamma}$	–	inner product in $H_0^1(\Omega)$,
		$\langle u,v\rangle_{H_0^1,\gamma} = \int_\Omega \nabla u \nabla v + \gamma uv\, dx,\ \gamma > 0$
$\|\cdot\|_{H_0^1,\gamma}$	–	corresponding norm in $H_0^1(\Omega)$,
		$\|u\|_{H_0^1,\gamma}^2 = \int_\Omega (\nabla u)^2 + \gamma u^2\, dx,\ \gamma > 0$
$\langle\cdot,\cdot\rangle_{H_0^2}$	–	inner product in $H_0^2(\Omega)$, $\langle u,v\rangle_{H_0^2} = \int_\Omega \Delta u \Delta v\, dx$
$\|\cdot\|_{H_0^2}$	–	corresponding norm in $H_0^2(\Omega)$, $\|u\|_{H_0^2} = \|\Delta u\|_{L_2}$
$\langle\cdot,\cdot\rangle_{H_0^2,\alpha}$	–	inner product in $H_0^2(\Omega)$,
		$\langle u,v\rangle_{H_0^2,\alpha} = \int_\Omega \Delta u \Delta v + \alpha uv\, dx,\ \alpha > 0$
$\|\cdot\|_{H_0^2,\alpha}$	–	corresponding norm in $H_0^2(\Omega)$,
		$\|u\|_{H_0^2,\alpha}^2 = \int_\Omega (\Delta u)^2 + \alpha u^2\, dx,\ \alpha > 0$
$H^{-m}(\Omega)$	–	dual space of $H_0^m(\Omega)$
$\|\cdot\|_{H^{-m}}$	–	usual operator norm in $H^{-m}(\Omega)$,
$\|\cdot\|_{H^{-2},\alpha}$	–	norm in $H^{-2}(\Omega)$ corresponding to the norm $\|\cdot\|_{H_0^2,\alpha}$ in $H_0^2(\Omega)$
E_X^Y	–	imbedding from the space X to the space Y
$C_{X\hookrightarrow Y}$	–	imbedding constant $X \hookrightarrow Y$

References

[1] Agmon, S., *The coerciveness problem for integro-differential forms*, J. Analyse Math., **6**, (1958), 183-223.

[2] Agmon, S., Douglis, A., Nirenberg, L., *Estimates near the boundary for solutions of elliptic partial differential equations satisfying general boundary conditions*, I. Comm. Pure Appl. Math., **12**, (1959), 623-727.

[3] Agmon, S., Douglis, A., Nirenberg, L., *Estimates near the boundary for solutions of elliptic partial differential equations satisfying general boundary conditions*, II. Comm. Pure Appl. Math., **17**, (1964), 35-92.

[4] Bandle, C., Reichel, W., *Solutions of quasilinear second-Order elliptic boundary value problems via degree theory*, in Handbook of Differential Equations, Stationary Partial Differential Equations, Volume 1, Chapter 1, edited by M. Chipot and P. Quittner, Elsevier B.V. (2004)

[5] Breuer, B., Horák, J., McKenna, P. J., Plum, M., *A computer-assisted existence and multiplicity proof for travelling waves in a nonlinearly supported beam*, J. Differential Equations, **224**, (2006), 60-97.

[6] Deimling, K., *Nonlinear functional analysis*, Springer-Verlag, 1985

[7] Douglis, A., Nirenberg, L., *Interior estimates for elliptic systems of partial differential equations* , Comm. Pure Appl. Math., **8**, (1955), 503-538.

[8] Engels, H., *Numerical quadrature and cubature* Academic Press, London, New York, Toronto, Sydney, San Francisco, 1980

[9] Friedrichs, K. O., *On certain inequalities and characteristic value problems for analytic functions and for functions of two variables*, Trans. Amer. Math. Soc. **41**, 321-364 (1937)

[10] Girault, V., Raviart, P.-A., *Finite element approximation of the Navier-Stokes equations*, Springer-Verlag, Berlin–Heidelberg–New York, 1979

[11] Hales, T. C., *Formal proof*, Notices of the American Mathematical Society,1370-1382, 2008

[12] Hammer, R., Hocks, M.,Kulisch, U., Ratz, D.,*C++ Toolbox for verified computing*, Springer-Verlag, Heidelberg, New York, 1995

[13] Hoang, V., Plum, M., Wieners, C., *A computer-assisted proof for photonic band gaps*, Z. angew. Math. Phys., **60**, (2009), 1035-1052.

[14] Horgan, C. O., Payne, L.E., *On inequalities of Korn, Friedrichs and Babuška-Aziz*, Arch. Rational Mech. Anal., **82**,(1983), no. 2, 165-179.

[15] Lehmann, N.J., *Optimale Eigenwerteinschlieungen*, Numer. Math.,**5**,(1963), 246-272.

[16] Moser, J.,*A sharp form of an inequality by N. Trudinger*, Indiana Univ. Math. J., **20**, (1971), 1077-1092.

[17] Nakao, M. T., *Solving nonlinear elliptic problems with result verification using an H^{-1} type residual iteration*, Computing Suppl. **9** (1993), 161-173.

[18] Nakao, M. T., Yamamoto, N., *Numerical verifications for solutions to elliptic equations using residual iterations with higher order finite elements*, Computing Suppl. **9** (1993), 161-173.

[19] Nirenberg, L., *Topics in nonlinear functional analysis*, Lecture Notes, vols. 1973-1974, Courant Institute of Mathematical Sciences, New York University, New York, 1974.

[20] Plum, M., *Explicit H_2-estimates and pointwise bounds for solutions of second-order elliptic boundary value problems*, J. Math. Anal. Appl. **165** (1992), 36-61.

[21] Plum, M., *Enclosures for weak solutions of nonlinear elliptic boundary value problems*, WSSIAA **3** (1994), 505-521.

[22] Plum, M., *Guaranteed numerical bounds for eigenvalues*, in Spectral Theory and Computational Methods of Sturm-Liouville Problems, Lecture Notes in Pure and Applied Mathematics Series/191, Marcel Dekker, Inc., New York, Basel, 1997

[23] Plum, M., *Computer-assisted enclosure methods for elliptic differential equations*, Lecture Course, Univ. Karlsruhe, Fak. für Mathematik, (2005)

[24] Plum, M., *A mixed-type explicit imbedding theorem and L_∞-error bounds for elliptic boundary value problems*, (2006), manuscript

[25] Plum, M., *Existence and multiplicity proofs for semilinear elliptic boundary value problems by computer assistance*, Jahresbericht der DMV, JB 110. Band (2008), Heft, 19-54.

[26] Plum, M., Wieners, C., *New solutions of the Gelfand problem*, J. Math. Anal. Appl. **269** (2002), 588-606.

[27] Reichel, W., Weth, T., *Existence of solutions to nonlinear, subcritical higher order elliptic Dirichlet problems*, Journal of Differential Equations, **248**, (2010), 1866-1878.

[28] Stoyan, G., *Towards discrete Velte decomposition and narrow bounds for inf-sup constants*, Comput. Math. Appl., **38**, (1999), 243261.

[29] Struwe, M., *Variational methods: Applications to nonlinear partial differential equations and Hamiltonian systems*, A Series of Modern Surveys in Mathematics, Springer-Verlag, Berlin, Heidelberg, 2008.

[30] Wieners, C., *Numerische Existenzbeweise fr schwache Lsungen nichtlinearer elliptischer Randwertaufgaben*, Dissertation, Kln, 1994

[31] Wieners, C., *Bounds for the N lowest eigenvalues of fourth order boundary value problems*, Computing **59**, (1997), 29-41.

[32] Wieners, C., *Parallel cubic finite elements on triangular meshes and smooth domains*, Script for a course in the summer school "hp finite elements", Karlsruhe, (2004)

[33] *InstitutsCluster User Guide, Version 0.93*, Karlsruhe Institute of Technology, Steinbuch Centre for Computing, 2009, http://www.scc.kit.edu/downloads/sca/ugic.pdf

[34] *HP XC3000 User Guide, Version 1.07*, Karlsruhe Institute of Technology, Steinbuch Centre for Computing, 2011, http://www.scc.kit.edu/scc/docs/HP-XC/ug/ug3k.pdf

MIX
Papier aus verantwortungsvollen Quellen
Paper from responsible sources
FSC® C105338

Printed by Books on Demand GmbH, Norderstedt / Germany